The Ecclesiology of Karl Rahner

The Ecclesiology of Karl Rahner

RICHARD LENNAN

CLARENDON PRESS · OXFORD
1995

Oxford University Press, Walton Street, Oxford OX2 6DP
Oxford New York
Athens Auckland Bangkok Bombay
Calcutta Cape Town Dar es Salaam Delhi
Florence Hong Kong Istanbul Karachi
Kuala Lumpur Madras Madrid Melbourne
Mexico City Nairobi Paris Singapore
Taipei Tokyo Toronto
and associated companies in
Berlin Ibadan

Oxford is a trade mark of Oxford University Press

Published in the United States
by Oxford University Press Inc., New York

British Library Cataloguing in Publication Data
Data available

Library of Congress Cataloging in Publication Data
The ecclesiology of Karl Rahner / Richard Lennan.
Includes bibliographical references and index.
1. Rahner, Karl, 1904– . 2. Church—History of doctrines—20th century. 3. Catholic Church—Doctrines. I. Title.
BX4705.R287L46 1995 262′.02′092—dc20 94–36057
ISBN 0–19–826358–9

1 3 5 7 9 10 8 6 4 2

Typeset by Graphicraft Typesetters Ltd., Hong Kong
Printed in Great Britain on acid-free paper by
Bookcraft (Bath) Ltd., Midsomer Norton

ACKNOWLEDGEMENTS

BOOKS not only tell a story, they also have their own story, a story of those who have guided, challenged, and nurtured the author. Accordingly, I am happy to have the opportunity to thank the people who have been part of the story of this book.

First, I would like to acknowledge the contribution of George Vass and Edward Yarnold, who supervised my work in Innsbruck and Oxford respectively. The perceptive criticism and warm encouragement I received from both George and Ted were a blessing I shall always treasure. I was also blessed in Roman Siebenrock of the Karl Rahner Archive in Innsbruck; his generosity and enthusiasm greatly facilitated my research. The project of research and writing was further enriched by the many conversations—and pizzas—I shared with Philip Endean, John Moffatt, and Patrick O'Liddy.

Secondly, there are many people who contributed to this work through their support for me rather than their knowledge of Rahner. In both England and Austria, I enjoyed much kindness and friendship. I am grateful particularly to Agnes and Raimund Eberharter, Ansgar Schocke, Norbert Rasim, Jim Lawlor, the Benham family, and the community of Campion Hall in Oxford and the Canisianum in Innsbruck. From the other side of the world, my mother and family, Lex Levey, Mark Lane, Gerard Bowen, Anthony Critchley, the Lynch family, and other friends made sure that I was neither too homesick nor too out of touch with life in Australia—and especially the fate of Canterbury-Bankstown in the Sydney Rugby League competition. Their support was, and is, much appreciated. I am also very grateful to my bishop, Leo Clarke, who made it possible for me to undertake doctoral work and who generously supported my study.

Thirdly, I would like to express my thanks to the students and staff of both the Catholic Institute of Sydney and St Patrick's College, Manly, especially Janiene Wilson and the third-year seminarians in 1994, for their friendship during the time I have spent revising the book for publication. The process of revision was also aided by Darryl Mackie, who saved me from disasters with my computer, Richard McMahon, who read the manuscript for me and has been a great source of encouragement, and Hilary O'Shea and her colleagues at Oxford University Press, whose kindness and professionalism have been a wonderful gift.

Finally, I would like to thank Karl Rahner himself. Although I did not have the opportunity to meet Rahner, his faith, freedom, and faithfulness have long inspired and challenged me. For both the inspiration and the challenge, I am deeply grateful.

I hope that all those whose kindness has helped to shape this book will accept it as an expression of my thanks.

R.L.

Excerpts from *Theological Investigations* reprinted by permission of:

Darton, Longman & Todd Ltd., London
The CROSSROAD Publishing Co., New York

CONTENTS

ABBREVIATIONS

AAS	*Acta Apostolicae Sedis*
AG	*Ad Gentes Divinitus*
DS	Denzinger–Schönmetzer, *Enchiridion Symbolorum*
DV	*Dei Verbum*
GS	*Gaudium et Spes*
LG	*Lumen Gentium*
NA	*Nostra Aetate*
UR	*Unitatis Redintegratio*
ST	*Schriften zur Theologie*
TI	*Theological Investigations*
TS	*Theological Studies*
ZKTh	*Zeitschrift für Katholische Theologie*

REFERENCES AND TRANSLATIONS

ALL references to Rahner's own works cite both a published English translation, where this exists, and the German original. To facilitate identification of any articles from *Theological Investigations* which are referred to more than once in the book, the first reference to such articles gives both the full title and, in square brackets, an abbreviated form of the title. This practice is also followed with the titles of many of Rahner's books.

In many cases, a published English translation has been slightly altered to allow either for more inclusive language or a more accurate rendering of the original. Where no English version exists, I have made my own translations.

There are realities which we understand only when we love them. The Church is one of these.

Karl Rahner

INTRODUCTION

EACH of the cliffs framing Whitby harbour in north Yorkshire is home to a distinctive monument. On one of the brows stands a statue of Captain James Cook, the master navigator of the eighteenth century, who not only served his apprenticeship in Whitby, but also sailed the locally built *Endeavour*, an undistinguished converted collier, to *terra australis incognita*. By way of contrast, the opposite promontory is commanded by the ruins of Whitby Abbey, a site which evokes Hilda, Cuthbert, and Bede, and the Synod which adopted Roman Christianity as the norm for Britain. The physical separation between Cook's statue and the Abbey's ruins parallels the gap between the world-views they represent.

Cook, a worthy exemplar of the Enlightenment spirit which esteemed discovery and progress, is the classic explorer. His journeys without maps opened for Europeans the way to a world whose very existence had been disputed by his contemporaries. While his achievements attest to the harvest only the audacious can reap, Whitby Abbey, the embodiment of the richness of Catholic tradition, proclaims that conserving what has been inherited is both a fundamental aspect of being human and a very particular feature of Christian faith. For this reason, even the ruins of the Abbey, the relics of a millennium of faithful witness to the Gospel, are cherished. The two guardians of Whitby harbour provide, therefore, an icon of the tension between the conflicting claims of progress and preservation.

While this tension is not of recent origin, it has acquired an increasingly bitter ideological dimension in recent decades. This is so because 'change', perhaps as a result of the staggering advances in technology since the Industrial Revolution, has come to imply neither evolution nor development, but a disjunction between the old and the new. Indeed, it could be argued that the modern world seems allergic to permanence and continuity. Nor has this notion of change as revolution been limited solely to science and industry; social and economic policies too have been the vehicles for dramatic, if not always beneficial, breaks with the past—witness, for example, the repercussions of 'Thatcherism' and 'Reaganomics'. Since contemporary proponents of change tend to exempt nothing from their vision of a new order, their reforming zeal usually evokes an equally uncompromising approach from their opponents. Hence the polarization.

Within the Church too, the last generation has witnessed struggles between advocates of radical change and supporters of the *status quo*, struggles which have been waged with an unrelenting ferocity. While divisions within the Church often seem to be simply another instance of that chasm between 'progressive' and 'conservative' world-views which manifests itself on issues as diverse as sport and architecture, they have, in fact, a unique character. To understand the specificity of intra-Church disputes, it must be remembered that until recent times it was changelessness which was popularly regarded as one of the Church's hallmarks.

Whether the notion of an unchanging Church is theologically defensible—or ever reflected a historical reality—is not as important as the fact that it has proved attractive. Indeed, the belief that the strength of the Catholic Church lay in its unchangeableness not only swayed Catholics, but, as is clear from the final section of Macaulay's laudatory, if florid, panegyric, also found adherents outside Catholicism:

> She saw the commencement of all the governments and of all the ecclesiastical establishments that now exist in the world; and we feel no assurance that she is not destined to see the end of them all. She was great and respected before the Saxon had set foot in Britain, before the Frank had passed the Rhine, when Grecian eloquence still flourished at Antioch, when idols were still worshipped in the temple of Mecca. And she may still exist in undiminished vigour when some traveller from New Zealand shall, in the midst of a vast solitude, take his stand on a broken arch of London Bridge to sketch the ruins of St. Paul's.[1]

Underlying the stress on the unchangeable nature of the Church, and so fuelling the vehemence of those opposed to change, was the identification of the Church as not just another human institution but a God-given aid to salvation. Consequently, it is little cause for wonderment that proposals which can be construed as

[1] Thomas B. Macaulay, 'An Essay on van Ranke's "History of the Popes"', in id., *Critical and Historical Essays Contributed to the Edinburgh Review* (London, 1870), 548.

tampering with the Church's beliefs, structures, or practices acquire a sensitivity unrivalled even by the ubiquitous controversies surrounding efforts to preserve historic buildings. While the latter can be reduced to the clash of aesthetic or historical judgements, suggestions for wide-ranging reforms in the Church can be seen as substituting human values for God's will and, consequently, as endangering our prospects for salvation.

None the less, the appearance of titles such as *The Runaway Church* and *The Reshaping of Catholicism* among contemporary works on ecclesiology is a clear indication that this notion of an unchanging and unchangeable Catholic Church no longer commands the field.[2] Indeed, in the decades since the Second Vatican Council, the issue of 'change' in the Church has become a shibboleth dividing the Catholic community. The tensions of the post-conciliar period evoke the gap which separates Whitby harbour's sentinels: for some, it is the spirit of Captain Cook, the willingness to embrace the new, to venture into the unknown, which is all-important—'anyone who wants the Church to die out, to become the grave of God, must want it to remain as it is. Anyone who wants it to live as God's living congregation, must want it to change';[3] for others, however, the changes which have occurred since the Council are scarcely distinguishable from vandalism—'And now the universal Church is in ruins. In a spectacular act of public suicide it has destroyed its cult and the culture it had informed. The world view that created Europe and kept the Word incarnate is shattered'—and only what has been inherited, the spirit of Whitby Abbey, is accepted as authentically Catholic.[4]

The fact that proposals for change in the Church generate strong emotions ought not, however, to be a barrier to exploring some fundamental issues associated with change; indeed, it makes the need for such an investigation more urgent. Since neither an inflexible rejection of the new nor the readiness to jettison everything

[2] Peter Hebblethwaite, *The Runaway Church* (New York, 1975); Avery Dulles, *The Reshaping of Catholicism: Current Challenges in the Theology of Church* (San Francisco, 1988).

[3] Hans Küng, *Truthfulness: The Future of the Church* (London, 1968), 141. That the years have not diminished Küng's commitment to substantial reform is clearly revealed in his later work: *Reforming the Church Today*, trans. P. Heinegg, F. McDonagh, J. Maxwell, E. Quinn, A. Snidler (New York, 1990).

[4] Anne Roche Muggeridge, *The Desolate City: Revolution in the Catholic Church*, rev. edn. (San Francisco, 1990), 182.

which appears 'dated' is ultimately helpful, it is crucial to seek a better perspective on what change involves. In fact, before one adopts a resolute 'for' or 'against' position on specific proposals for change, several important questions concerning the nature of change need to be answered: Is change, irrespective of how much or how little, actually reconcilable with the Church's God-given nature? Does change necessarily imply a radical break with the past or is it consistent with continuity? What legitimates change in the Church? What are the limits of change? What is the relationship between change and the Church's existence in history? That, until recent times, there has been little interest in such questions is a further indication of the long-prevailing rejection of any possibility of change in the Church.

Indeed, according to the ecclesiastical historian John O'Malley, change in the Church has traditionally been understood exclusively in terms of restoring the Church to its original purity by reforming abuses. Until last century, argues O'Malley, the Church was dominated by the conviction that religion was to change people, not vice versa.[5] Such an attitude discouraged the development of a hermeneutic which emphasized the positive dimensions of change. In the last hundred years, however, a new awareness of the implications of historical existence has made it possible to substantiate the paradoxical claim that change, far from being subversive of identity, is actually the precondition for maintaining intact what is essential to the authenticity of that identity.[6]

As a result of his historical research, O'Malley has developed a three-tiered categorization, which attempts both to delineate the processes by which change in the Church takes place and to indicate that change is a far more nuanced term than is popularly supposed. According to O'Malley, all changes in the Church can be grouped under one or other of the following captions: *developments*, which can be of greater or lesser intensity and involve changes in mentalities or structures—such developments occur not as the result of deliberate initiatives by the leaders of the

[5] John W. O'Malley, 'Historical Thought and the Reform Crisis of the Sixteenth Century', *TS* 28 (1967), 536.

[6] This understanding of change comes from O'Malley's 'Developments, Reforms, and Two Great Reformations: Towards an Historical Assessment of Vatican II', *TS* 44 (1983), 374. For a review of the way change has been understood in the history of the Church, see O'Malley's, 'Reform, Historical Consciousness and Vatican II's Aggiornamento', *TS* 32 (1971), 591–5.

Church, but because of 'outside influences', such as the impact on the Church of Hellenism, of printing in the Middle Ages, and of radio and television in our own time; *reformations*, where members of the Church initiate changes in the Church's self-consciousness for the presumed good of the Church—in terms made popular by Thomas Kuhn, such reformations, among which O'Malley includes the medieval Investiture Controversy and the Lutheran Reformation, necessitate a 'paradigm shift', a new way of understanding the Church; *reforms*, which are 'adjustments' or 'emendations' within a system, but which do not give rise to the need for a new paradigm.[7]

By building on O'Malley's analysis of the processes of change, Avery Dulles has devised his own three-tiered approach to 're-form'—a word he understands as indicating a level of change which keeps intact the identity of what is being changed. Dulles's primary concern is with the 'ends' of change. Thus, he argues that change can have one of three aims: first, the restoration of an original state of perfection—the traditional understanding of change in the Church; secondly, the actualization of potentialities—in this category Dulles places Newman's interpretation of development as an organic process;[8] thirdly, expediting a more creative interaction of the Church with society. This third model conceives of the Church not as a 'substance' with a fixed and immutable nature, but as being formed by its relationship with society. Consequently, the Church develops in response to shifts in the human environment of which it is a part.[9]

Taken together, the insights of O'Malley and Dulles lead to the conclusion that 'change' is far more of a variegated process than many contemporary progressives or conservatives allow. The work of O'Malley and Dulles also provides valuable conceptual tools

[7] O'Malley, 'Developments', 375–6. O'Malley's discussion of the Investiture Controversy and the Reformation can be found at pp. 380–90. Kuhn's notion of a 'paradigm shift', and particularly its application to theology, is discussed in greater detail by Hans Küng in Küng and David Tracy (eds.), *Paradigm Change in Theology*, trans. M. Kohl (Edinburgh, 1989), 7–10.

[8] The classic statement of Newman's approach is, *An Essay on the Development of Christian Doctrine* (Westminster, Md., 1968), which he wrote in 1845, while still an Anglican. For a discussion of Newman's understanding of development, see Aidan Nichols, *From Newman to Congar: The Idea of Doctrinal Development from the Victorians to the Second Vatican Council* (Edinburgh, 1990), 17–70.

[9] Avery Dulles, *The Resilient Church: The Necessity and Limits of Adaptation* (Dublin, 1978), 31–3.

for assessing the implications of specific proposals for change in the Church. In the present work, this apparatus will be applied to the ecclesiology of Karl Rahner.

Before making some introductory comments on Rahner's ecclesiology, it is worth considering why any aspect of Rahner's theology ought to be studied. This question is especially pertinent today, when a concentration on hermeneutics, language, and methodology may result in Rahner's theology, concerned as it is with the more 'applied' issues of faith—such as the believer's relationship to the Church or the Church's relationship to the world—being regarded as vaguely old-fashioned. Such a prejudice does less than justice to Rahner. Indeed, the attraction of Rahner remains his conviction, which will be highlighted more than once in this book, that theology was not 'art for art's sake', but was legitimate only if it addressed itself to the real concerns of people in their particular period of history.

While Rahner's claim that his bibliography was more dramatic than his biography, might suggest that Rahner can have been little more than a cloistered academic, this impression is not accurate.[10] Certainly, Rahner was far from being a swashbuckling action-hero. His life was, however, more interesting than may have been anticipated by his birth in 1904 into a 'normal, middle-class, Christian family' of seven children—one of whom, his elder brother Hugo, also made a significant contribution to contemporary theology as a patristic and Ignatian scholar—or his entry into the Jesuits when he was only 18.[11] By the time of his death in 1984, a few weeks after his eightieth birthday, Rahner had not only completed sixty years as a Jesuit and fifty years as a priest, but he had also been either a participant in, or an observer of, trends and events which reshaped not only theology and the Church, but the wider society.[12]

[10] For Rahner's attitude towards both his biography and biblography, see *I Remember*, trans. H. Egan (London, 1985), 58–9 (*Erinnerungen* (Freiburg i.B., 1984), 62–3).

[11] Ibid. 24 (ibid. 21).

[12] Among the numerous studies of Rahner's work, there are many which detail both his biography and the influences on his intellectual development. See e.g.: William V. Dych, *Karl Rahner* (London, 1992), 4–17; Geffrey B. Kelly (ed.), *Karl Rahner: Theologian of the Graced Search for Meaning* (Minneapolis, 1992), 1–31; Robert Kress, *A Rahner Handbook* (Atlanta, 1982), 1–18, 62–70; Herbert Vorgrimler, *Understanding Karl Rahner: An Introduction to his Life and Thought*, trans. J. Bowden (New York, 1986), *passim*; see also the autobiographical interview, *I Remember (Erinnerungen)*, *passim*.

Rahner was, for example, at the forefront of the movement away from the rarefied scholaticism which had dominated Catholic theology in the first half of the twentieth century. His theology was influenced not only by the Fathers and Thomas Aquinas, but also by his contemporaries, like Joseph Maréchal and, most importantly, Martin Heidegger—in whose seminars Rahner had participated while a doctoral student at the University of Freiburg im Breisgau—whose primary focus was on the experiencing subject, not abstract essences. During his long teaching-career—spent largely in Innsbruck, but also with shorter stints in Munich, where he succeeded Romano Guardini, and Münster—Rahner developed a theology characterized by its emphasis on identifying God as central to all human experience, rather than to the narrowly 'religious' sphere of life.

Accordingly, and this will be a major concern of this book, Rahner was passionate in his conviction that the Church needed to be aware of, and to respond to, developments in history and culture. Failure to do so meant that the Church would end by preaching a God irrelevant to the real concerns of humanity. Significantly, Rahner's conviction on this point was not the product of a theoretical construct, but grew out of witnessing the atrophy of 'the Christian West' in his own time. Indeed, as will be noted in Chapter 4, the demise of the Christian culture which had dominated much of European history radicalized Rahner's theology of the Church. Rahner's conviction that theology had to be done in a new key often resulted in developments that made his biography as dramatic as his bibliography.

Thus, Rahner's original doctoral thesis was rejected because it was regarded as too influenced by Heidegger's thought—the failed thesis was later published as *Spirit in the World* and remains a landmark in the history of contemporary Catholic theology. Rahner ultimately obtained his doctorate from the University of Innsbruck, for a thesis on the patristic interpretation of John 19: 34. Not long after completing his doctoral and *Habilitation* studies, Rahner, together with the rest of the Jesuit community, was expelled from the Tyrol when the Nazis dissolved the theology faculty in Innsbruck. The war years, most of which he spent in Vienna, were also an important period in the evolution of Rahner's ideas on Christianity's relationship to the contemporary world—Rahner's most significant writing of the war years, the 'Vienna Memorandum', will be discussed in Chapter 4.

Rahner's difficulties with those who were unappreciative of his approach to theology were not exhausted by his doctoral examiner in Freiburg i.B. Thus, on more than one occasion before the Second Vatican Council, Rahner's work was subject to censorship or other penalties because it was out of step with the prevailing ecclesiastical atmosphere. Vatican II, however, was the catalyst of Rahner's 'rehabilitation'. This was so both because of his involvement in the Council as a *peritus*, and because the Council's commitment to renewing the relationship between the Church and the world reflected Rahner's own commitment to developing a theology attuned to the needs of its own time. This commitment also provided the rationale for Rahner's discussions, in the years after the Council, with philosophers influenced by Communism, and other non-traditional dialogue partners—a development which paralleled his discussions with Protestant theologians during the years before Vatican II, when Catholic–Protestant relations were cool, if not positively frosty. Such discussions highlighted his belief that the Church could not remain aloof from any aspects of its cultural context.

Perhaps because of the momentous events he witnessed, Rahner's work is characterized by a commitment to the future. The future which he envisaged was not, however, to be simply the continuation of the present. It challenged the Church, therefore, to find new ways of communicating the richness of the Christian tradition. As a corollary of his conviction that the Christian tradition had to be open to the future, Rahner's numerous publications on ecclesiology—which spanned a period of fifty years—called on the Church to be open to change in its doctrines, structures, in its relationship with the world, and in the relationship between different groups within the Church itself.

Since Rahner's ideas on how the Church might change to respond to contemporary challenges and opportunities constitute a major focus of this book, it is appropriate that one of the first tasks of the book will be to indicate that his ecclesiology never lacked an understanding of the Church's need and capacity for change. As will be seen, Rahner's understanding of change in the early period of his work had the following characteristics: first, he envisaged the Church changing in the direction of becoming more itself, of growing to be what it was always meant to be—Dulles's organic model; secondly, it was Rahner's conviction that the impetus for such change was provided by the Spirit, who was both the source of the Church's development and the bane of any tendency in the Church to ossification; thirdly, Rahner was committed to the view that change was fully reconcilable with the maintenance of continuity.

A further object of the book is to illustrate that Rahner's understanding of the processes of change shifted, that there are differing emphases between the 'earlier' and 'later' Rahner. While there was always a tension in Rahner's work between, to use Dulles's categories, the organic approach to change and the promotion of change on the basis of the Church's relationship to a evolving world, it will be argued here that it was the latter which gradually became the dominant feature of his understanding of change in the Church. As will be illustrated, this shift in emphasis was occasioned by his assessment of the challenge which the social conditions of the second half of the twentieth century posed to the Church. In addition, the pastoral dimension of Rahner's theology was also far more to the fore in his later work, particularly in his proposals for the future of the Church.

The goal of demonstrating a development in Rahner's own understanding of change in the Church has determined the method of this present work. What follows, therefore, is essentially a chronological exposition of his ecclesiology. Accordingly, it begins by examining the foundations of Rahner's approach to the Church, foundations laid principally in a variety of articles written in the 1940s and 1950s. The aim in so doing is to show that an organic understanding of change—that is, the notion of change as the development of potentialities—was integral to Rahner's early view of the Church. In the second half of the book, the emphasis is on Rahner's conviction that contemporary social conditions had become the primary force for change in the Church. Consequently, Part Two begins by focusing on Rahner's analysis of the challenges to the Church—and, therefore, the pressure for change—arising from the complexity of life in the twentieth century. Following this, his response to Vatican II and his proposals for the future shape of the Church—both of which represented his attempt to respond to contemporary social conditions—will be explored.

The final section of the book will assess whether Rahner's specific proposals for the future were consistent both with his own understanding of the nature of the Church and with the maintenance of continuity in the Church. Thus, what will be explored is whether the later Rahner was, to use O'Malley's terms, advocating reform—a level of change which maintains continuity with what has gone before—or reformation—a paradigm shift which establishes a decisive break with the past. In addition the Conclusion will also discuss whether Rahner's understanding of change can help to bridge the gap between the progressives and conservatives in the modern Church.

Before undertaking a detailed analysis of Rahner's ecclesiology, two further points of clarification about this book are needed. The first concerns the meaning of the word 'Church'. While much that Rahner wrote is applicable to all Christian churches, it is none the less true, as will be seen at the beginning of Chapter 1, that Rahner in his early work unequivocally identified 'Church' as indicating the 'Roman Catholic Church' and always regarded himself as a Catholic theologian. At the same time, however, as will also be indicated, Rahner was committed to the oneness of the Church, a commitment sustained throughout his life. Consequently, while 'Church' in this work is synonymous with the Catholic Church, this usage is not designed to be sectarian, but to indicate what Rahner desired for the Church as a whole.

The second point concerns how Rahner's theological method is discussed in this book. Although it has become conventional in Rahner studies to begin with an exposition of his ontology and 'transcendental method' before proceeding to discuss other aspects of his work, this convention will not be observed here. Instead, what will be attempted—principally in the first three chapters—is to situate Rahner's ontology and theological method within the context of his ecclesiology. Thus, such major Rahnerian themes as the 'supernatural existential', the *Vorgriff*, and the relationship between the 'categorial' and the 'transcendental' will be examined via his theology of the symbol, revelation, Christology, and the work of the Holy Spirit in the Church, rather than as isolated themes. This strategy has been adopted in the belief that the various aspects of Rahner's methodology can best be understood by examining their application. More specifically, what this approach suggests is that the full import of Rahner's methodology can most clearly be perceived in his ecclesiology—the subject-area to which more than half of his writings are devoted. Finally, since the Conclusion will be dedicated to assessing the overall value of Rahner's ecclesiology, specific criticism of either his method or his particular proposals for change will be examined at the end of each chapter.

PART ONE

1

The Church: Sacrament of Christ

In a simpler age, Humpty Dumpty could inform Alice that words meant whatever he wanted them to mean. In our own time, however, Ludwig Wittgenstein and other linguistic philosophers have made us aware that communication, the end at which language aims, is possible only when meaning is shared. Consequently, before beginning to expound on Rahner's theology of change in the Church, it is necessary to detail how he understood 'Church'. As will be seen, Rahner himself not only provided such a description, but was also anxious to show that his ecclesiology was not simply idiosyncratic. In order to illustrate that his theology was other than a gnosticism to which only the illuminati could have access, Rahner made use of that control mechanism which can be found exclusively in Christianity: the teaching of the Church.

Rahner's emphasis on the indispensable role of Church teaching appeared as early as his 1947 commentary on the theme of 'membership of the Church' in *Mystici Corporis Christi*. In that article, which remained his most substantial ecclesiological text in the period prior to Vatican II, Rahner claimed that it was integral to the identity of Catholic theology that the doctrinal pronouncements of the Church should constitute its beginning and end.[1] In adopting this hermeneutical principle, Rahner was not implying, however, that the theologian could do no more than defend and comment on, rather than develop, the doctrine of the teaching authority. Indeed, he specifically disavowed the suggestion that the only authentic Catholic theology was the so-called 'Denzinger theology'—a pejorative expression implying that much masquerading as creative theology is actually little more than a collection

[1] 'Membership of the Church According to the Teaching of Pius XII's Encyclical "Mystici Corporis Christi"' ['Membership'], *TI* ii. 2 ('Die Gliedschaft in der Kirche nach der Lehre der Enzyklika Pius XII "Mystici Corporis Christi"', *ST* ii. 8).

of references to conciliar texts.[2] None the less, the notion of the 'end' did suggest that the teaching of the Church always remained 'an abiding norm' for the work of the theologian.[3] How Rahner understood the dynamics of such a norm will be a major concern of this work.

Far from believing that faithfulness to an ecclesial tradition produced a theology distinguished only by being insipid, Rahner maintained the real enemy of creative theology was a lack of regard for tradition. Accordingly, he argued that a series of logical deductions from an arbitrarily chosen notion of 'Church' could never rival the richness already present in the Church's self-understanding.[4] As interpreted by Rahner, faithfulness to the teaching authority neither acted as the bane of creativity nor rendered theologians incapable of addressing the needs of their contemporaries. A prime example of how Rahner himself made use of official teaching to develop an ecclesiology which did not merely clone Council or encyclical is his response to the definition of 'Church' in *Mystici Corporis*.

'CHURCH' IN MYSTICI CORPORIS CHRISTI

From *Mystici Corporis*, Rahner distilled that 'Church' denominated: 'the Roman Catholic Church, which knows itself to be founded by Christ, even as an external, visibly organised society with the Bishop of Rome as its head, and which as such declares itself to be basically necessary for salvation'.[5]

Basing itself on this understanding, the encyclical taught that only those who, in addition to having been validly baptized, professed the true faith and were subject to the Church's powers of order and jurisdiction could be classed as members of the Church.[6] Although Rahner's response to this definition of membership will not be discussed until later in the chapter, what is immediately relevant is his ready acknowledgement of the importance of the visible Church. Indeed, he claimed that any attempt to substitute some sense of merely inner belonging

[2] 'Membership of the Church According to the Teaching of Pius XII's Encyclical "Mystici Corporis Christi"' ['Membership'], n. 2 ('Die Gliedschaft in der Kirche nach der Lehre der Enzyklika Pius XII "Mystici Corporis Christi"', n. 1).

[3] Ibid. (ibid.). The ref. is to the same footnote.

[4] Ibid. 4 (ibid. 10). [5] Ibid. 5 (ibid. 40). [6] Ibid. 27 (ibid. 33).

contradicts in its very deepest sense the incarnational principle of Christianity, according to which God was made flesh and grace attached to the concrete, historical here and now of human realities. And according to this principle, God has not left it to the free choice of human beings to decide for themselves in what concrete form and historically verifiable reality they wish to find Christ's salvation and the grace of God.[7]

While he thus affirmed the basic thrust of *Mystici Corporis*, Rahner also recognized that the emphases of the encyclical were not merely fortuitous. What the doctrine of *Mystici Corporis* revealed was that the teaching authority of the Church, that authority which, as Rahner himself acknowledged, had the right to determine how words were to be used in the context of expressing the truths of faith, wanted 'Church' to be understood in a specific way at a particular point of history—the idea of Church authorities overseeing how words were used will be discussed in detail in a later chapter.[8] Accordingly, Rahner contended that the stress in *Mystici Corporis* on the visible reality of the Church was to be interpreted in light of the encyclical's *Sitz im Leben*. The 'external, visibly organised society with the Bishop of Rome as its head' was highlighted as a response to the view, common in ecumenical discussions of the time, that even those separated from one another in belief could still form one Church, if 'Church' was understood as an 'inward pneumatic community of love'.[9]

Although Rahner did not suggest that either its origin or its accent negated the worth of the encyclical, he was certainly convinced that its deliberately narrow perspective disqualified *Mystici Corporis* from claiming to be the ultimate ecclesiological statement. Indeed, he feared that unless balanced by other considerations, the document's emphasis on the Church as a visible communion could be responsible not merely for an impoverished sense of 'Church', but for an 'ecclesiological Nestorianism' which ignored everything but the visible.[10] That *Mystici Corporis* was itself aware of such a danger was not denied by Rahner; indeed, he accepted that the encyclical had recognized that portraying the

[7] Ibid. 34 (ibid. 40). [8] Ibid. 5 (ibid. 11).

[9] Ibid. 34 (ibid. 40). For the background to the encyclical, see H.-J. Schulz, 'Kirchenzugehörigkeit', in Elmar Klinger and Klaus Wittstadt (eds.), *Glaube im Prozeß* (Freiburg i.B., 1984), 397–8.

[10] 'Membership', *TI* ii. 70 (*ST* ii. 77). As early as 1943, in his unpub. 'Vienna Memorandum'—which will be discussed in detail in Ch. 4—Rahner had already warned of the dangers of 'ecclesiological Nestorianism'.

Church as merely a juridical and social organization would be a rationalistic error.[11] Where Rahner went beyond the encyclical, however, was in devising a more comprehensive ecclesiology.

In other words, while accepting the doctrine of the encyclical as an 'abiding norm' which was not to be contradicted, Rahner proceeded to develop an ecclesiology which emphasized not only the Church's historical form, but also such aspects as its existence as the Body of Christ, animated and sanctified by the Spirit.[12] In so doing, he saw himself fulfilling what he regarded as the proper mission of the theologian: to combine the various strands of Christian tradition into a logical framework.[13] In formulating his ecclesiology, Rahner's aim was to navigate the shoals between the Scylla of a Nestorian approach and the Charybdis of portraying the Church as the invisible community of those sharing a particular religious feeling. The key to achieving this balance lay in what became his trade-mark image of the Church: the Church as a sacrament. Central to this notion of the sacramentality of the Church was his understanding of the nature of a symbol.

SACRAMENT AND SYMBOL

In order to appreciate better the details of Rahner's notion of the Church as a sacrament, there is a need to review how sacramentality in general has traditionally been understood in the Catholic context. This can best be done by beginning with the Council of Trent, which codified the Church's doctrine on the sacraments.

As expressed by Trent's teaching on the Eucharist, a sacrament 'symbolum esse rei sacrae et invisibilis gratiae formam visibilem'.[14] What is crucial here is that sacraments are understood neither as merely internal and invisible encounters between the grace of God and the human recipient nor simply as external signs, but rather as events of grace taking place under and through the *symbolum*. For Rahner too, the 'inner' and 'outer' aspects of a sacrament were equally indispensable to any authentically Catholic notion of sacramentality:

[11] 'Membership', *TI* ii. 70 (*ST* ii. 77).
[12] Ibid. 71 (ibid.). [13] Ibid. 3 (ibid. 8–9).
[14] This ref. to the Council of Trent's decree on the Eucharist can be found in DS 1639.

One can give a definition of a Sacrament which includes exclusively only what belongs to the constitution of a *valid* sacramental sign. And, on the other hand, one can give a description of the nature of a Sacrament which also includes that reality in its notion for which the external sacramental event is essentially intended and without which it would ultimately lose its meaning. Both these notions of Sacrament, if we may put it this way, are necessary and indispensable, neither can replace the other; they must not be played off one against the other.[15]

A sacrament, therefore, could not be considered to have been fully defined if the definition alluded only to the visible (*sacramentum*). In order for the visible to be properly understood, it had to be seen as a symbol of a deeper reality (*res sacramenti*), which was not visible. This did not mean, however, that what was visible was not essential. Indeed, the sacramental signs themselves—or, more correctly, symbols—precisely as signs, were the causes of grace.[16] To make sense of this seeming conundrum, what needs to be investigated is Rahner's notion of the symbol.

In an article written in 1959, an article which Hugo Rahner saw as embodying an essential element of his brother's theological approach to the question of how God could be experienced in the world, Rahner sought to make precise the meaning of 'symbol'.[17] Accordingly, he defined a symbol as that through which a being expressed itself in a way necessary for its own self-realization.[18] Furthermore, in order to emphasize that, far from being simply

[15] 'Membership', *TI* ii. 72 (*ST* ii. 79). Rahner's emphasis.

[16] *The Church and the Sacraments* [*Church and Sacraments*], trans. W. J. O'Hara (London, 1963), 37 (*Kirche und Sakramente* [*Sakramente*] (Freiburg i.B., 1960), 34).

[17] Hugo's comment, which occurs in an open letter he wrote to Karl on the occasion of the latter's sixtieth birthday, can be found in Johann Baptist Metz (ed.), *Gott in Welt*, ii (Freiburg i.B., 1964), 897.

[18] 'The Theology of the Symbol' ['Symbol'], *TI* iv. 234. ('Zur Theologie des Symbols', *ST* iv. 290. The definition in the original runs thus: 'Das eigentliche Symbol (Realsymbol) ist der zur Wesenskonstitution gehörende Selbstvollzug eines Seienden im anderen.') Rahner's theology of the symbol is critically assessed by James Buckley in 'On Being a Symbol: An Appraisal of Karl Rahner', *TS* 40 (1979), 453–73; for other views on Rahner's understanding of symbols, see: C. Annice Callahan, 'Karl Rahner's Theology of the Symbol: Basis for his Theology of the Church and the Sacraments', *Irish Theological Quarterly*, 49 (1982), 195–205; Michael J. Walsh, *The Heart of Christ in the Writings of Karl Rahner: An Investigation of its Christological Foundation as an Example of the Relationship between Theology and Spirituality* (Rome, 1977), 19–33; Nancy Clasby, 'Dancing Sophia: Rahner's Theology of Symbols', *Religion and Literature*, 25 (1993), 51–65; the development of Rahner's understanding of symbols is discussed by Joseph H. P. Wong in *Logos-Symbol in the Christology of Karl Rahner* (Rome, 1984), 39–73.

an image or a likeness of something else, a symbol actually existed in a 'differentiated unity' with what it symbolized, he referred not merely to a symbol, but to a symbolic reality or *Realsymbol*.[19] Rahner's insistence that symbols not be confused with lesser comparative terms reflected his conviction that symbols are a foundational element of human existence. Indeed, he argued that all beings, because they necessarily 'express' themselves in order to attain their own nature, were symbolic.[20] In other words, it was true both that no being could be understood without reference to a symbol, and that what appeared externally did not exhaust all that could be known of any being. Thus, the physical object which could be experienced immediately—the 'categorial'—was to be understood as the symbol of a deeper reality—the 'transcendental'.[21] This distinction between the categorial and the transcendental was of fundamental importance for Rahner's theology of revelation, which will be examined in the next chapter.

For Rahner, the connection between the symbol and what was symbolized was not accidental but intrinsic and mutual: not only was the symbol the means by which another reality became present in the world, but what was symbolized determined the way the symbol itself was understood. In order to explain how this could be so, Rahner drew on Thomistic ontology, specifically on the notion of formal causality.

As Rahner interpreted Thomas, it was the giving of the form by the formal cause which actually brought about what was formed. In the context of the sacraments, for example, this meant that grace—which, as will be shown, Rahner understood as God's self-communication—actually produced the sacramental sign.[22] While

[19] 'Symbol', *TI* iv. 234 (*ST* iv. 290). For a discussion of the notion of a *Realsymbol*, see Wong, *Logos-Symbol*, 75–82.

[20] 'Symbol', *TI* iv. 224 (*ST* iv. 278). The Ger. orig. is: 'das Seiende ist von sich selbst her notwendig symbolisch, weil es sich notwendig "ausdrückt", um sein eigenes Wesen zu finden'. Rahner's discussion of how it is possible for one thing to be the symbol of another can be found at pp. 222–35 (276–91).

[21] Ibid. 229–31 (ibid. 283–86). The development and philosophical ancestry of Rahner's distinction between the appearance and the deeper reality is discussed by L. B. Puntel in 'Zu den Begriffen "tranzendental" und "kategorial" bei Karl Rahner', in Herbert Vorgrimler (ed.), *Wagnis Theologie: Erfahrungen mit der Theologie Karl Rahners* (Freiburg i.B., 1979). 189–98; see also James J. Bacik, *Apologetics and the Eclipse of Mystery: Mystagogy According to Karl Rahner* (Notre Dame, Ind., 1980), 22–7.

[22] 'Symbol', *TI* iv. 232 (*ST* iv. 287).

this external manifestation of the form was different from the form itself, it nevertheless was to be understood as a symbol of that form—the external manifestation did not, therefore, have an existence independent of the formal cause.[23] Thus, although the symbol and what was symbolized were not the same, they were inseparable. For this reason, the symbol could be distinguished from a mere sign, which did not bear an intrinsic relationship to what it signified.[24]

While this concept of the symbol resists casual analysis, it was, however, no mere excursion into abstract ontology. To illustrate its concrete importance, and simultaneously complete the infrastructure which buttresses the concept of the Church as a sacrament, it is necessary to focus on how Rahner's ideas on symbols can be applied to the paradigm of sacramentality: Christology.

CHRIST: SACRAMENT OF THE FATHER AND FULLNESS OF HUMANITY

Both prior to and since Chalcedon, the primary task of Christology has been to deepen our understanding of John 1: 14: the Word became flesh. For Rahner, the theology of the Logos was a theology of the symbol. Put simply, the humanity of Jesus was a symbol for his divinity. Jesus's humanity was the 'appearance' of the Logos itself, its symbolic reality (*Realsymbol*) in a pre-eminent sense.[25] As such, it was not something alien to the Logos, not something taken up from outside which did not reveal anything about the one who made use of it, but was, in a way consistent with the approach to formal causality discussed above, the self-disclosure of the Logos itself. In short, when God chose to be revealed in the Logos, what appeared was the humanity of Christ: 'The Logos, as son of the Father, is truly in his humanity as such, the revelatory symbol in which the Father enunciates himself, in this Son, to the world—revelatory because the symbol renders present what is revealed.'[26] Expressed in these terms, Rahner's

[23] Ibid. (ibid. 288). [24] Ibid. 225 (ibid. 279).

[25] Ibid. 238 (ibid. 295). For a discussion of the use of symbol in Rahner's Christology, see: Wong, *Logos-Symbol*, 113–84; Annice Callahan, *Karl Rahner's Spirituality of the Pierced Heart: A Reinterpretation of Devotion to the Sacred Heart* (Lanham, Md., 1985), 32–53; and Walsh, *The Heart of Christ*, 97–100.

[26] 'Symbol', *TI* iv. 239 (*ST* iv. 296).

Christology evokes a crucial question: how is it possible for finite humanity to be a symbol of infinite divinity? To appreciate Rahner's answer to this question, it is necessary to combine his approach to formal causality with both his interpretation of grace and his theological anthropology.

As early as 1939, in an article on scholastic notions of grace, Rahner had stressed that God's relationship to human beings was to be understood in terms of formal causality, not efficient causality. While the latter would have implied merely an accidental modification from outside of an already-complete human nature, Rahner argued that both the hypostatic union and the bestowal of grace involved a 'being taken up into the ground' of the form itself.[27] This understanding of humanity in terms of God's formal causality enabled Rahner to draw two conclusions which were of fundamental importance for his whole theological project: grace was to be defined not as a property of God, but as God's self-communication (this concept, which was central to Rahner's theology of revelation, will be expounded in the next chapter); and openness to God was constitutive of human existence.

In the 1950s, in a series of articles which represented a response to the views of Henri de Lubac and other practitioners of *nouvelle théologie*, Rahner expressed his distinctive understanding of the relationship between nature and grace.[28] Central to the theology he developed in these articles was the concept of 'the supernatural existential', a vehicle for interpreting human existence in terms of humanity's God-given capacity to receive God: 'The capacity for the God of self-bestowing personal Love is the central and abiding existential of human beings as they really are.'[29] By using the

[27] 'Some Implications of the Scholastic Concept of Uncreated Grace', *TI* i. 329 ('Zur scholastischen Begrifflichkeit der ungeschaffenen Gnade', *ST* i. 358).

[28] For the background to the debate—and a discussion of Rahner's response—see Dych, *Karl Rahner*, 32–48. Further analysis of Rahner's theology of grace can be found in: Stephen J. Duffy, *The Graced Horizon: Nature and Grace in Modern Catholic Thought* (Collegeville, Minn., 1992), 85–114; Roger Haight, *The Experience and Language of Grace* (Dublin, 1979), 119–43; and George Vandervelde, 'The Grammar of Grace: Karl Rahner as a Watershed in Contemporary Theology', *TS* 49 (1988), 445–59.

[29] 'Concerning the Relationship between Nature and Grace' ['Relationship between Nature and Grace'], *TI* i. 312 ('Über das Verhältnis von Natur und Gnade', *ST* i. 339), orig. pub. 1950. Rahner's other articles from the 1950s which combine grace, anthropology, and Christology are: 'The Eternal Significance of the Humanity of Jesus for our Relationship with God' ['Humanity of Jesus'], *TI* iii. 35–46 ('Die ewige Bedeutung der Menschheit Jesu für unser Gottesverhältnis', *ST* iii.

'supernatural existential' to summarize the relationship between grace and nature, Rahner was able both to protect the sovereign freedom of God and to show that openness to God was indeed the 'central and abiding' feature of being human.[30] If it was God's desire to communicate love which brought humanity into being—'God wishes to communicate God's self, to pour forth the love which God is. . . . And so God makes a creature whom God can love: God creates humanity'[31]—not only was this God's free choice, it also meant that human beings could not be understood without reference to grace, to God's self-communication.

Accordingly, grace was neither extrinsic to human nature—which could have suggested that human beings were ultimately independent of God—nor intrinsic to it—which could have suggested that grace was a necessary component of the human make-up, rather than God's gift. To avoid both of these flawed approaches, Rahner argued that 'pure nature' was merely a 'remainder concept' (*Restbegriff*), that which would be left if the supernatural existential were removed.[32] In reality, however, the fact that God's life was offered to all meant that no human beings ever existed in a state of pure nature.

Just as Rahner's Christology established the need for an anthropology, so too the anthropology aided the articulation of the Christology. In other words, Rahner's Christology combined both 'descending' and 'ascending' elements. Thus, he portrayed Christ not simply as the sacrament of God, but also as the fullness of humanity. If, because of God's self-communication, the human being was 'a reality absolutely open upwards', then the highest perfection of humanity, the realization of humanity's highest possibility, was when the Logos became present in the world in that humanity—this point will be discussed further in the next chapter,

47–60), pub. 1953; 'Reflections on the Experience of Grace', *TI* iii. 86–90 ('Über die Erfahrung der Gnade', *ST* iii. 105–9), pub. 1954; 'Thoughts on the Theology of Christmas', *TI* iii. 24–34 ('Zur Theologie der Weihnachtsfeier', *ST* iii. 35–46), pub. 1955; 'On the Theology of the Incarnation', *TI* iv. 105–20 ('Zur Theologie der Menschwerdung', *ST* iv. 137–55), pub. 1958; and, 'Nature and Grace', *TI* iv. 165–88 ('Natur und Gnade', *ST* 209–36), pub. 1959.

[30] The value of, and challenges to, the supernatural existential are assessed by George Vass, *Understanding Karl Rahner*, ii: *The Mystery of Man and the Foundations of a Theological System* (London, 1985), 64–83.

[31] 'Relationship between Nature and Grace', *TI* i. 310 (*ST* i. 336–7).

[32] Ibid. 313 (ibid. 340).

in connection with Rahner's theology of revelation.[33] Christ was, therefore, both the fullness of humanity and the primordial sacramental word of God, the word which not only spoke of the love and mercy of God, but actually made them present in his person.[34] Christ was the sacrament of God's commitment to humanity, that commitment whose irrevocability and invincibility stemmed not from human acceptance, but from God's free decision.[35] This symbolic or sacramental approach to Christology found its echo in Rahner's ecclesiology.

THE SACRAMENTALITY OF THE CHURCH

As Christ was the sacrament of the Father, so the Church was the sacrament of Christ himself. Expressed in terms of sacramental theology, the Church was

> the abiding presence of that primordial sacramental word of definitive grace, which Christ is in the world, effecting what is uttered by uttering it in sign. By the very fact of being in that way the enduring presence of Christ in the world, the Church is truly the fundamental sacrament, the well-spring of the sacraments in the strict sense. From Christ, the Church has an intrinsically sacramental structure. Historically visible in space and time with its double aspect as people of God and as juridical and social organisation of this people, the Church is the body and bride of Christ who abides in the Church as the presence in the world of God's historical and eschatological self-promise during this last of its epochs. God does not abandon the Church and cannot do so, since God wills to remain forever in the flesh of the one human family.[36]

Although the assistance of his theology of the symbol was required to bring to maturity Rahner's notion of the Church as a sacrament, the idea had actually been gestating since the 1930s. That the Church was neither merely an exclusively internal nor merely an exclusively external fellowship was a conviction of Rahner's long before he took up the cudgels against the one-dimensional approach evident in *Mystici Corporis*. Indeed, as early as 1934 the sacramental model was adumbrated in his claim that

[33] 'Current Problems in Christology', *TI* i. 183 ('Probleme der Christologie von heute', *ST* i. 204), orig. pub. 1954.
[34] *Church and Sacraments*, 18 (*Sakramente*, 17).
[35] Ibid. (ibid.). [36] Ibid. (ibid.).

The supernatural life of redeemed humanity becomes visible in the historical uniqueness of the here and now of the Church on this earth, just as it also entered historically into humanity by revelation. Thus the supernatural life—which at least in itself, seems to lie quite apart from the human, historical reality—appears supported by what is visible and human, brought down into the earthly point of time, dependent on worldly things.[37]

Furthermore, in 1942, the year before the publication of *Mystici Corporis*, Rahner had specifically emphasized the Christological dimension of the Church's sacramentality:

If the 'Church' is then, *before* being the visible social organisation—even though this is its necessary expression—the social accessibility of the historico-sacramental permanent presence of the salvation reality of Christ, a society founded in its essential features directly by Christ himself, it follows that those who hold offices within that visible social organisation of the Church do not first create the 'Church', that is the possibility of universal historical mediation, but presuppose it.[38]

While his appreciation of the workings of grace in the world gradually broadened to encompass those outside the Church, the period before Vatican II witnessed no equivocation in Rahner's emphasis on the unity of the Church's 'interior' and 'exterior' aspects. What remained constant, therefore, was his conviction that the Church could best be understood as 'the union of the interior graced relationship of the redeemed and the historical, visible form of this transcendent interior union'.[39]

At no stage, however, did Rahner veer towards ascribing divinity to the Church. He did not, therefore, refer to the Church in a way which implied that its manifest historical form and the Holy Spirit were one and the same. Nevertheless, it remained true that if the Church was indeed the sacrament of Christ, the means by which his grace was embodied in history, then its historical form and the Spirit could no more be separated than could the divinity

[37] 'The Meaning of Frequent Confessions of Devotion' ['Confessions'], *TI* iii. 184 ('Vom Sinn der häufigen Andachtsbeichte', *ST* iii. 219).

[38] 'Priestly Existence', *TI* iii. 248 ('Priesterliche Existenz', *ST* iii. 295). Rahner's emphasis. The pub. Eng. trans. refers to the Church proceeding from 'the mere existence of Christ himself'; this is far more than implied by the original 'von Christus selbst'.

[39] 'Personal and Sacramental Piety' ['Sacramental Piety'], *TI* ii. 122 ('Personale und sakramentale Frömmigkeit', *ST* ii. 129–30), orig. pub. 1952.

of Christ be divorced from his humanity.[40] Consequently, Rahner could claim that a rejection of the ecclesial character of grace was also a rejection of the relationship between grace and the Incarnation.[41]

That the Church existed as the symbol of Christ—'the persisting presence of the incarnate word in space and time'—also sufficed to guarantee its permanence.[42] If the Church could have ceased to exist, it would have implied that what God had achieved in Christ—'the historically real and actual presence of the eschatological mercy of God'—was, in reality, less than final.[43] The presence of Christ in the Church not only meant, therefore, that the Church had a future, it also determined the quality of that future: because of Christ, the Church could never become a meaningless symbol.[44] Unlike the Synagogue, the Church would always remain the definitive symbol of grace—'the Church of the last days, the final and irrevocable economy of salvation, the concrete representation of God's free, triumphant, unconditional design for our salvation'.[45] Since the Church was the symbol of God's offer of grace, of God's self, in history, it followed that its actions were the means by which this grace was imparted. Foremost among such actions was the celebration of the sacraments.

Fundamental to Rahner's theology of the sacraments was the principle that their value and meaning could be understood only when they were perceived as aspects of the Church, 'the primordial sacrament'.[46] Indeed, he argued that as the existence of the

[40] *Church and Sacraments*, 19 (*Sakramente*, 18).
[41] Ibid. 22 (ibid. 20). [42] 'Symbol', *TI* iv. 240 (*ST* iv. 297).
[43] *Church and Sacraments*, 14 (*Sakramente*, 13).
[44] Ibid. 18–19 (ibid. 17–18).
[45] *Inspiration in the Bible* [*Inspiration*], trans. H. Henkey (rev. trans. M. Palmer) (New York, 1964), 40–1 (*Über die Schriftinspiration* [*Schriftinspiration*] (Freiburg i.B., 1958), 47).
[46] 'Symbol', *TI* iv. 241 (*ST* iv. 298). The description of the Church as the 'primordial' or *Ursakrament* became prominent through the work of Otto Semmelroth, *Die Kirche als Ursakrament* (Frankfurt am Main, 1953). Semmelroth's approach is summarized in Bernard Przewozny, *The Church as the Sacrament of the Unity of All Mankind* (Rome, 1979), 41–6. The history of the notion of the Church as sacrament is expounded by Leonardo Boff, *Die Kirche als Sakrament* (Paderborn, 1972); Boff discusses Rahner's theology on pp. 314–22. That this notion of the Church as the primordial sacrament is not an attempt by Rahner to relegate Christ to a secondary position is made clear in *Church and Sacraments*, 19 (*Sakramente*, 18), where Rahner explains that the Church is the primordial sacrament only in relation to the ecclesial sacraments, not to Christ himself.

Church symbolized the grace of Christ present in history, the seven ecclesial sacraments in their turn made concrete and actual for the life of the individual the symbolic reality of the Church itself.[47] Consequently, it was through the sacraments that the Church attained 'the highest degree of actualization of what it always is: the presence of redemptive grace for humanity in the historical tangibility of her appearance, which is the sign of the eschatologically victorious grace of God in the world'.[48]

Since the sacraments thus symbolized the grace of Christ at work through the Church, they were more than human constructs. To illustrate this, Rahner stressed that the Church had not abstractly deduced the seven sacraments from an a priori understanding of its own essence, but had in fact come to recognize that essence only in its concrete fulfilment in the sacraments.[49] In other words, the Church's awareness that it was offering redemptive grace through the sacraments enabled it to perceive that certain actions were expressions of its own deepest reality. It was these actions which were then classed as sacraments. The fact that the sacraments thus shared in the nature of the Church—in both its 'internal' and 'external' aspects—meant that they, no less than the Church itself, could never be deprived of meaning. Their integrity derived, however, not from the personal holiness of those who composed the Church, a holiness which—as will be seen in the following section—was vulnerable to human weakness, but from Christ, whose grace they symbolized.[50]

As a result of the Church's link to Christ, it followed that the mere celebration of the sacraments within the Church—*opus operatum*—sufficed to effect the offer of grace, the offer of God's self, to those who received them.[51] As with membership of the Church itself, however, this did not imply that it was possible to speak of a 'physical certainty of functioning' in regard to grace.[52] Even though the offer of grace through the Church's ministration of the sacraments could be unequivocally affirmed, grace was not a magical process which functioned independently of the disposition of the recipient. Consequently, the efficacy of grace, its acceptance by those to whom it was offered, remained inscrutable

[47] 'Symbol', *TI* iv. 241 (*ST* iv. 199).
[48] *Church and Sacraments*, 22 (*Sakramente*, 21).
[49] Ibid. 70 (ibid. 63).
[50] Ibid. 28–9 (ibid. 26).
[51] Ibid. 28 (ibid. 25).
[52] Ibid. 26 (ibid. 24).

to all but God. Properly understood, therefore, the sacraments involved the mysterious interaction of God and the believer in the context of the Church:

> A sacrament takes place . . . as a dialogical unity of the personal acts of God and of the person in the visible sphere of the Church's essential (that is, given to it directly by Christ himself) sanctifying, official action. In the sacraments, the incarnation and process of becoming historically tangible of grace reach their high-point.[53]

While the emphasis on the indissoluble link between the Church and the conferral of grace clearly establishes the centrality of the Church to Christian life, it is not without its dangers. Indeed, in his assessment of sacramentality as a 'model' for understanding the Church, Avery Dulles noted that it can induce a 'narcisstic aestheticism' which naïvely glories in the Church as the earthly home of the Spirit.[54] Such an attitude would imply that the Church had already achieved perfection, that it is divine. In the context of a Church requiring neither development nor reform, it would, therefore, be meaningless to talk of 'change'. As will be seen in the next section, however, Rahner's approach to the Church's sacramentality successfully avoided such a trap.

THE SINFUL, HOLY CHURCH

In the years before Vatican II, much of the literature in ecclesiology separated sinful members from the Church in order to highlight the latter's sinlessness.[55] Rahner, however, pursued a different tack. Indeed, in his earliest foray into this field, an article written in 1934, he acknowledged both that sinners remained members of the Church—a key consideration in avoiding an idealized Church—and that the Church was indeed affected by such sinfulness.[56] None the less, what was not present in this early contribution was an explicit confession that the Church itself must therefore be regarded as sinful.

When, more than a decade later, Rahner returned to this theme,

[53] 'Sacramental Piety', *TI* ii. 125 (*ST* ii. 132–3).

[54] Avery Dulles, *Models of the Church* (New York, 1978), 78.

[55] For a brief survey of pre-conciliar ecclesiology, see Avery Dulles, 'A Half Century of Ecclesiology', TS 50 (1989), 419–21.

[56] 'Confessions', *TI* iii. 187 (*ST* iii. 223).

an evolution in his approach was evident. This development manifested itself in his rejection of the view that, although the members of the Church could be sinful, the Church itself was incorruptibly holy. Drawing on the sacramental model, Rahner argued that the ineradicable union of the visible and invisible dimensions of the Church meant that if the members of the Church were sinful, then the Church itself had also to be considered sinful, since it did not exist independently of those who formed it.[57] The sinfulness of the Church was, therefore, not simply a fact of experience, but a truth of faith. Although sinners, like valid but unfruitful sacraments, lacked a life-giving relationship to the Spirit, they did not thereby forfeit their membership of the Church.[58] As a result, the Church itself was sinful.

Since even membership of the Church's hierarchy did not confer immunity from flawed human nature, actions initiated in the name of the Church by members of the hierarchy could not be protected against the effects of sin. Indeed, Rahner emphasized that the Spirit was not obliged to ensure that the sinfulness of the Church's office-holders would have an impact only in their private lives.[59] At the same time, however, he also claimed that members of the Church could refuse obedience to such authorities only when commanded to do something objectively sinful. The belief that a particular directive from a superior was motivated by a sinful narrowness was not itself sufficient justification for a refusal to obey.[60]

For Rahner, it was not enough simply to establish that the Church was sinful; he was also anxious to identify the source of such sinfulness. Thus, in an article written immediately prior to the opening of Vatican II, he contended that it was the Church's unwillingness to accept that the Spirit would guarantee its future which was the core of its sinfulness. This lack of trust often manifested itself in a refusal to change:

> it is the Church which does not have the courage to regard the future as belonging to God in the same way as it has experienced the past as belonging to God. The Church is often in the position of one who glorifies the past and looks askance at the present in so far as it has not

[57] 'The Church of Sinners' ['Sinners'], *TI* vi. 259 ('Kirche der Sünder', *ST* vi. 308), orig. pub. 1947.

[58] Ibid. (ibid.). [59] Ibid. 261 (ibid. 310). [60] Ibid. (ibid.).

created the present itself, finding it all too easy to condemn it. . . . The Church has quite often in the past sided with the powerful and made itself too little the advocate of the poor. . . . It often places more value on the bureaucratic apparatus of the Church than in the enthusiasm of the Spirit; it often loves the calm more than the storm, the old which has proved itself more than the new which is bold.[61]

In contrast to the earlier articles on this theme, this later piece—perhaps reflecting Rahner's own difficulties with the Holy Office in the decade before the Council[62]—was far more direct in identifying how the Church's sinfulness had expressed itself in history. Significantly, the emphasis was again on the Church's fear of the future:

Often in the past, the Church has in its office-bearers wronged saints, thinkers, those who were painfully looking for an answer, or theologians—all of whom merely wanted to give it their selfless service. Often before, it has warded off public opinion in the Church, although according to Pius XII, such public opinion is essential to the well-being of the Church. Not infrequently it has mistaken the barren mediocrity of an average theology and philosophy for the clarity of a good scholastic tradition.[63]

In his 1947 article on sin in the Church, Rahner had also discussed the challenge which this sinfulness posed to the believer. Far from suggesting that sinfulness—of its members in general and of office-holders in particular—meant that God had abandoned the Church, Rahner affirmed that even the sinful Church remained 'the bride of Christ and the vessel of the Holy Spirit'.[64] The Church's sinfulness proclaimed neither that God had forsaken it nor that it had become merely another ultimately corruptible human institution, but that the working of God's grace in it was a mystery.[65] Consequently, his guiding principle was that people of faith ought neither to rejoice in nor to condemn the sinful Church, but to recognize that they too were sinners whose

[61] 'Thoughts on the Possibility of Belief Today' ['Belief Today'], *TI* v. 16 ('Über die Möglichkeit des Glaubens heute', *ST* v. 25), orig. pub. 1962.

[62] For Rahner's own description of these difficulties, see *I Remember*, 63–5 (*Erinnerungen*, 68–70). A more complete exposition of the background to Rahner's problems can be found in Karl-Heinz Neufeld, 'Lehramtliche Mißverständnisse: Zu Schwierigkeiten Karl Rahners in Rom', *ZKTh* 111 (1989), 420–30.

[63] 'Belief Today', *TI* v. 16 (*ST* v. 25–6).

[64] 'Sinners', *TI* vi. 261 (*ST* vi. 310).

[65] 'Membership', *TI* ii. 76 (*ST* ii. 83).

sinfulness had obscured the light of the Gospel.[66] The most creative response, therefore, was not only to refrain from rejecting the Church, but also to hope that the Church would be led through its sinfulness to a deeper conversion. While the Church's sinfulness was a burden for its members, Rahner was emphatic that the Church's sacramental structure precluded the possibility of separating its corrupt human elements from the immaculate Spirit of God. As a result, there could be no flight from the wounded Church of history to an ideal one:

> every spirituality, no matter how deep, which can no longer bear both the figures of the Church as virgin and as woman of sin—in humility and love, with the forbearance and patience of God—reveals itself after a short period as fanaticism, as the spectre of a spirituality where human beings in the end remain caught up in themselves.[67]

Nevertheless, Rahner stressed that sin in the Church, just as in the life of the individual, was never anything other than a contradiction of its deepest reality. Although sin could not be extirpated from the Church's historical form, it did not exist in order that God's grace might be more abundantly revealed, but actually veiled what only holiness could fully disclose.[68] Indeed, he argued that since its sinfulness distorted the presence of Christ in the world, the Church could at times make it more difficult for people to discover Christ.[69]

While he clearly accepted that it could properly be described as sinful, Rahner nevertheless stressed that the Church was not a pure paradox of the union of visible sin and invisible grace. Thus, even the sinful Church did not cease to proclaim, through the indwelling Spirit, the holiness of God. Similarly, he was unwavering in his commitment to the view that it was holiness, since it reflected the presence of Christ, which was the quintessential expression of what the Church would infallibly and indestructibly remain.[70]

Despite the fact that human sinfulness meant that the victory of God's justifying grace would never become fully apprehensible in history, this sin could not supplant grace. As a result, even the sinful human being—and, *a fortiori*, the Church—was fundamentally a

[66] 'Sinners', *TI* vi. 267 (*ST* vi. 318).
[67] Ibid. 266 (ibid. 315).
[68] Ibid. 263 (ibid. 313).
[69] Ibid. 262 (ibid. 312).
[70] Ibid. 263 (ibid. 313).

new creation in Christ. If the individual and the Church remained *simul iustus et peccator*, it was only in the sense that sin in the present masked what would inevitably be the final victory of grace.[71]

The reality of sin notwithstanding, Rahner claimed that the holiness of the members of the Church, which could be empirically ascertained by the person of good will enlightened by faith, still made the Church a motive for faith and witnessed to its divine commission.[72] Consequently, it was the saints, the embodiment of the Church's holiness, who made a peerless contribution to revealing the true nature of the Church.[73] Rahner argued that the canonization of saints did not merely proclaim that God's grace was made accessible through the Church, but that, as manifested in their lives, this grace was also efficacious. It was, therefore, through the saints that the Church actually became in fact what it was in theory: *signum levatum in nationes*.[74] Consequently, the saints, 'the cloud of witnesses', did more than fulfil a human need for heroes: they revealed that the Church was the community of eschatological salvation and victorious grace.[75]

The impact of grace and the witness of the saints meant that it was neither a falsification of the facts nor a glossing-over of hard truths to present the history of the Church as other than a catalogue of scandals.[76] Rahner stressed that the achievements of the Spirit of God in the Church remained more important than the record of human meanness. For this reason

> the truest history of the Church . . . would be the history of the saints. Not only of those who have been canonised, but of all those in whom there has really taken place this miracle of pneumatic existence as the discovery of grace-given individuality in a selfless opening of the innermost kernel of the person's being towards God and so towards all spiritual persons.[77]

[71] 'Justified and Sinner at the Same Time', *TI* vi. 221–2 ('Gerecht und Sünder zugleich', *ST* vi. 265–6), orig. pub. 1963.

[72] 'Sinners', *TI* vi. 262 (*ST* vi. 311).

[73] 'The Church of the Saints' ['Saints'], *TI* iii. 93–4 ('Die Kirche der Heiligen', *ST* iii. 113–14), orig. pub. 1955.

[74] Ibid. 95 (ibid. 115). The ref. is to Vatican I's *Dei Filius* (DS 3014).

[75] 'Saints', *TI* iii. 96 (*ST* iii. 117).

[76] 'Sinners', *TI* vi. 267 (*ST* vi. 317).

[77] 'On the Significance in Redemptive History of the Individual Member of the Church' ['Individual Member'], in *Mission and Grace*, i [*Mission*], trans. C. Hastings (London, 1963), 139 ('Über die heilsgeschichtliche Bedeutung des einzelnen in der Kirche', in *Sendung und Gnade* [*Sendung*] (Innsbruck, 1959), 107). This was the orig. publication of the article.

As was clearly demonstrated by the ineradicable nature of its sinfulness, the fact that the Church was the sacrament of God's grace in history did not bestow divinity on it. Consequently, in discussing how the believer ought to view the Church, Rahner distinguished between the Church and the Kingdom of God present in its fullness. This did not imply, however, that he regarded the Church as being without salvific significance. In fact, with a particular emphasis from the beginning of the 1960s, Rahner attempted to establish a mean between an exaggerated 'ecclesiological piety' and a total lack of appreciation for the Church's role in the economy of salvation. How he did so, and what his approach reveals about the possibility of change in the Church, will be examined in the next two sections.

THE CHURCH: AN OBJECT OF FAITH?

That Christian faith proclaimed that the Parousia alone would occasion the fulfilment of both individuals and the Church itself confirmed for Rahner the Church's status as a product, rather than the source, of salvation.[78] On the other hand, he stressed that its existence as the sacrament of Christ's presence in history identified the Church as more than a human response to grace. Although it was not the source of salvation, the Church was certainly the channel through which salvation was offered to the world. This was so since the Spirit of Christ, whose becoming 'all in all' would constitute the Parousia, was already present in the Church.[79] Consequently, the Church could be described as:

> The community of those who already possess the eschatological gift which is God, who in full liberty accept this possession, who confess in faith that the possession and its acceptance have been caused by the free action of God's love and who hope for the unveiling of the possession by the power contained within it. The Church belongs therefore both to the present time in so far as the Church is still moving towards the goal by faith and hope, that is in so far as it must still let the Parousia of Christ come upon it, and at the same time the Church belongs to eternity as the Church moves towards its end in virtue of the future which has already arrived.[80]

[78] 'The Church and the Parousia of Christ' ['Parousia'], *TI* vi. 297 ('Kirche und Parousie Christi', *ST* vi. 350), orig. pub. 1963.

[79] Ibid. 299 (ibid. 352). [80] Ibid. 304 (ibid. 358).

Nevertheless, the fact that the Church too was awaiting the consummation of the Kingdom, proved that it had not yet attained perfection. For this reason, Rahner cautioned against an exclusive portrayal of the Church as the 'teaching Church', whose faith was complete. What could not be overlooked, was the Church's need to grow in faith. The Church did not, therefore, simply dispel the darkness for its followers, but, like them, had to endure that darkness with faith in God's providence.[81] In short, Rahner rejected any suggestion that the Church, unlike its members, was other than a pilgrim. In addition, the Church's pilgrim state meant that, the presence of the Spirit notwithstanding, it was not above either criticism or reform:

If the Church can never end up outside the truth of Christ, does that also mean that the Church proclaims this truth with that strength, with that topicality and always newly appropriated form which would make it salutary and which one might long for? Is it really always and clearly the case that by transforming this truth, by opening it into the infinity of God, by comforting and saving with it, that the Church allows it to become most intimately united with all the boundless, wild, confusing and yet so glorious chaos of perceptions, questions, notions, intellectual conquests, unfathomable perplexities, which we call the 'world picture', the world view, of modern humanity? Is not the permanence of the Gospel message in the Church often purchased (contrary to the meaning of Gospel truth) at the price of guarding fearfully against exposing ourselves to this 'chaos' (out of which the world of tomorrow will be born) or, at best, by meeting it purely defensively, trying merely to preserve what we have?[82]

Although Rahner's detailed critique of the Church's response to the modern world will not be examined until Chapter 4, two implications of the above quote are worth noting immediately: first, because it was the sacrament of Christ, the Church was protected against ultimate apostasy; secondly, as was discussed in the previous section, its sacramental nature did not immunize the Church against either mediocrity or fear.

While Rahner acknowledged that the Church could properly be understood only via a 'total engagement' which recognized that it was more than a human construct, he was none the less insistent

[81] 'Dogmatic Notes on "Ecclesiological Piety"' ['Ecclesiological Piety'], *TI* v. 340–1 ('Dogmatische Randbemerkungen zur Kirchenfrömmigkeit', *ST* v. 383–4), orig. pub. 1961.

[82] Ibid. 339 (ibid. 382).

that the same claims could not be made for the Church as for God.[83] This meant, for example, that while it was possible to say not merely that God was an object of our faith, but that we believe in God, the Church could be described only as an object of faith.[84] The key to appreciating that this was not simply a semantic quibble lies in exploring Rahner's understanding of the act of faith.

As interpreted by Rahner, faith was a response to grace. It was so, as it involved not merely assent to propositions about God, but a relationship with God. As has already been discussed, Rahner believed that God's grace, God's self-communication, not only initiated this relationship, but even established the human capacity for it. Thus, in regard to God, it could be affirmed that what was believed—God—was, simultaneously, also the ground of being for the act of faith.[85] Faith, therefore, both began with God and led to God. What distinguished the Church from God was that the former, because its very existence depended on God, not on itself, was radically incapable of becoming the 'ground' of such an act of faith.

In addition, despite the fact that its sacramental nature meant that it was more than the sum of its individual members, the Church was not a human being, who could give or receive that personal surrender of which individuals were capable.[86] As this surrendering of oneself was epitomized in the act of faith, it followed that such faith could not be directed towards the Church. Similarly, love, respect, and loyalty—all of which were aspects of faith—could rightly be predicated only of persons. They could, therefore, be applied to the Church only in a derived sense or when exercised towards the people who made up the Church.[87]

Although the Church could not establish itself as the origin or end of faith, Rahner nevertheless recognized two ways in which it could properly be described as 'the ground of faith'. First, the Church, in being open to all, witnessed to the basic Christian belief that the saving word of God was intended not for a select few but for the whole of humanity.[88] As such, the Church embodied a further element of the nature of grace: its communal dimension. Rahner argued that grace reached people in so far as they

[83] Ibid. 336 (ibid. 379). [84] Ibid. 343 (ibid. 386).
[85] Ibid. 343–4 (ibid. 386–7). [86] Ibid. 349 (ibid. 393).
[87] Ibid. (ibid.). [88] Ibid. 344 (ibid. 387).

were members of the 'one community of damnation and salvation'.[89] The proof for this claim, suggested Rahner, was that the highest point of becoming a person—which, *ipso facto*, was coterminous with the most profound response to grace—was attached to a community of persons: the communion of saints.[90]

Secondly, the Church could be the ground of faith because it was the authoritative bearer of the salvific word of Christ.[91] Since faith involved the overcoming of self in order to enter into the mystery of God, a being possessed by God rather than possessing God, hearing the word of God from an authoritative source was integral to it.[92] Surrendering oneself to the more comprehensive faith of the community of the Church, rather than remaining imprisoned within one's own limited insights was, therefore, a constitutive element of faith—this notion will recur later, in the discussion of Rahner's emphasis on the link between truth and institution.

Rahner stressed, however, that even regarding the Church as an object of faith was possible only for those who first believed in God. Indeed, he argued that, unlike Augustine, who professed that he accepted the Gospel only because it was proclaimed by the authority of the Church, contemporary Christians were prepared to entrust themselves to the Church only because Christ had not only founded it, but passed on his authority to it.[93] This primordial faith alone, suggested Rahner, enabled believers to stand by the Church even when its relationship to Christ was obscured by human weakness.[94] Consequently, the real glory of the Church was not its visible 'glitter', which, although God-effected, was finite, but rather the faith of those who believed—*contra spem in spem*—that, God could work even through a sinful Church.[95]

If, without denying the distinction between God and the Church, it could thus validly be maintained that the Church was an indispensable support for faith, then it followed that membership of the Church was not without significance for salvation. How Rahner assessed the nature and limits of this significance will be explored in what follows.

[89] 'Sacramental Piety', *TI* ii. 122 (*ST* ii. 129).
[90] Ibid. (ibid.).
[91] 'Ecclesiological Piety', *TI* v. 345 (*ST* v. 388).
[92] Ibid. 345–6 (ibid. 388–9).
[93] Ibid. 352–3 (ibid. 396).
[94] Ibid. 351 (ibid. 394–5).
[95] Ibid. 342 (ibid. 385).

THE VALUE OF CHURCH MEMBERSHIP

The Church's existence as the fundamental sacrament of God's mercy meant that salvation—the triumph of this mercy in the life of the individual—was both offered and promised to those who entered into a positive relationship with the Church.[96] Rahner stressed, however, that while membership of the Church could thus be considered the basic sacramental sign of the sanctification of the individual, it remained known to God alone, as was the case with reception of all the sacraments, whether the grace offered through such membership was effective.[97] Membership of the Church was, therefore, not a magical invocation which bound God, but an aspect of the mystery of grace.

Although he thus emphasized that membership of the ecclesial community did not guarantee entry to heaven, Rahner was none the less insistent that the Church, as the sacrament of Christ, played a role in the transmission and increase of divinizing grace. It fulfilled this role through its guidance and good example, its communal prayer, and the sacraments.[98] Consequently, without denying that Catholics too had to work out their salvation in fear and trembling, Rahner was convinced that dying in the peace of the visible Church was an additional ground for hope in God's mercy.[99]

Rahner's efforts to highlight the value of Church membership were matched, however, by his concern to disparage the notion that the Church was the community of those with an exclusive claim to salvation. Thus, without departing from his belief that the Church was God's chosen means of grace in history, Rahner was insistent that those outside the Church were not necessarily 'lost'.[100]

This concern for those outside the Church had been apparent as early as his commentary on *Mystici Corporis*. Indeed, in that article, Rahner had argued that, when combined with the ancient

[96] '*Church and Sacraments*, 21 (*Sakramente*, 19).

[97] 'Membership', *TI* ii. 76 (*ST* ii. 83).

[98] 'Some Theses on Prayer "In the Name of the Church"', *TI* v. 426 ('Thesen über das Gebet "im Namen der Kirche"', *ST* v. 479), orig. pub. 1961.

[99] 'The Christian among Unbelieving Relations' ['Unbelieving Relations'], *TI* iii. 361–2 ('Der Christ und seine ungläubigen Verwandten', *ST* iii. 426–7), orig. pub. in 1954.

[100] Ibid. 362 (ibid. 427).

teaching *extra ecclesiam nulla salus*, an exclusive emphasis on juridical membership of the Church could imply that non-members were actually cut off from that grace which alone was the source of salvation.[101] Consequently, his response to *Mystici Corporis* was largely an attempt to show that those who did not satisfy the encyclical's criteria for being members of the Church could nevertheless be said to belong to the Church in 'lesser and looser ways'.[102] Such a strategy allowed Rahner to defend both the necessity of the church as a means of saving grace and the possibility that this grace was operative even in the lives of those who were not baptized Catholics enjoying communion with the Holy See.

For anyone aware of later developments in his writings, it will be clear that the argument outlined above is a prelude to what is perhaps both the most celebrated and the most controversial theme of the Rahnerian corpus: the theology of the 'anonymous Christian'.[103] In fact, this theory was essentially an attempt to demonstrate how the truth enshrined in *extra ecclesiam nulla salus*—this concentration on the core of truth in a dogma was to become an increasingly prominent feature of Rahner's theological method—could be maintained in a world where Christians were an ever-decreasing minority. Indeed, the idea of the 'anonymous Christian' acted as a counterpoint to Rahner's assessment of the changing position of the Church in an increasingly secularized society.

Thus, even before Vatican II addressed itself to the prospects of salvation for those outside the Church, Rahner was promoting the belief both in an explicit offer of grace in the Church, and in the 'anonymous' offer outside it. On the one hand, he affirmed that

[101] For a survey of the history and interpretation of *extra ecclesiam nulla salus*, see Hans Küng, *The Church*, trans, R. and R. Ockenden (London, 1967), 403–8.

[102] 'Membership', *TI* ii. 55 (*ST* ii. 62). See pp. 34–62 (German version pp. 40–68) for the arguments—including the distinction between 'Church' and *corpus Christi mysticum*—supporting his claim that even the unbaptized can be members of 'the Church'.

[103] For the seeds of 'anonymous Christianity' in Rahner's pre-Vatican II writings, see: 'Unbelieving Relations', *TI* iii. 355–72 (*ST* iii. 419–39); and the 1961 article, 'Christianity and the Non-Christian Religions', *TI* v. 115–34 ('Das Christentum und die nichtchristlichen Religionen', *ST* v. 136–58). The most detailed secondary study of the development and content of Rahner's theory is Nikolas Schwerdtfeger's *Gnade und Welt: Zum Grundgefüge von Karl Rahners Theorie der 'anonymen Christen'* (Freiburg i.B., 1982); more summary presentations can be found in Walter Kern, *Außerhalb der Kirche kein Heil*? (Freiburg i.B., 1979), 69–77, Kress, *A Rahner Handbook*, 56–61, and Francis A. Sullivan, *Salvation outside the Church? Tracing the History of the Catholic Response* (New York, 1992), 162–81.

the unbaptized lacked the 'objective aids' for salvation provided by the common life of Church members and the direction given by the Church itself.[104] On the other hand, however, he argued that if the Church was understood as equivalent to 'people of God', then even the unbaptized could, by virtue of a *votum ecclesiae*, be numbered among its members. The foundation of this latter claim—which even as sympathetic a critic as Avery Dulles refers to as 'somewhat idiosyncratic' and a less sympathetic critic, Medard Kehl, rejects as having no connection with the biblical notion of 'the people of God'[105]—was his view that the Incarnation had effectively forged the whole of humanity into 'the people of God'.[106]

Since the possibility of salvation both for those in the Church and for those outside it derived from the same source, Rahner interpreted the certainty of God's offer within the Church as the sacrament of what was actually offered to all:

> If the history of humanity is one in which everyone from Abel to the last person is connected and in which everyone means something for everyone else and not merely for those living at the same time and place on earth, then the Church is the leaven not only when we can see with our own eyes that it has taken hold of a part of the rest of the flour and has made itself part of the fermentation, but always and for each and every age and also where the flour has not (yet) changed itself into a leavened dough in a way which is tangible for us.[107]

Thus understood, the Church was God's promise to the world that salvation was a possibility for the whole of humanity. Since twentieth-century conditions actually made it increasingly unlikely that the whole of humanity would ever formally enter the Church, Rahner argued that fulfilment of this promise was not contingent on the Church achieving a paramount position in society.[108] Indeed, it was belief in this promise which was to sustain Christians even when they experienced the circumstances of the Church in the world as shrouded in darkness. The sacramental understanding of the Church thus provided not only an encouraging view of

[104] 'Membership', *TI* ii. 59 (*ST* ii. 65).

[105] Dulles, 'Half Century of Ecclesiology', 432; Medard Kehl, *Kirche als Institution: Zur theologischen Begründung des institutionellen Charakters der Kirche in der neuen deutschsprachigen katholischen Ekklesiologie* (Frankfurt am Main, 1976), 186.

[106] 'Membership', *TI* ii. 83–4 (*ST* ii. 89–90).

[107] 'Ecclesiological Piety', *TI* v. 361 (*ST* v. 406).

[108] Ibid. 362 (ibid.).

the prospects for salvation both of those within the Church and of non-Christians, but, as has already been indicated, the basis for trusting that God's providence would secure the future of the Church.

Not surprisingly, the positive note which characterized Rahner's assessment of the prospects for salvation of non-Christians was reflected in his approach to ecumenism. Indeed, as will be apparent in the next section, Rahner not only recognized the significance of the non-Catholic Christian churches, but was eager that Catholics and Protestants work together to promote the Gospel in a society which found it increasingly alien.

ECUMENISM

Even in the 1950s, when the 'cold war' between Catholics and Protestants was no less frosty than that between 'the East' and 'the West', Rahner emphasized that affirmation of the Catholic Church's special role in the economy of salvation did not imply that the Protestant Churches were therefore to be considered 'a religiously negative quantity'.[109] Indeed, he argued that since the latter had, in their theology, piety, liturgy, community life, and art, actualized genuine Christian potentialities which had not been realized within the Catholic Church, Protestantism was misrepresented if portrayed as merely a successive dilution of authentic Christianity.[110] While not disputing that there remained in Protestantism aspects of doctrine which, from a Catholic perspective, were erroneous, Rahner denied that Catholics had either the right or the duty to assert

> that everything 'over there' which is not simply inherited ready-made from the time before the division is false, nor do they have to act as though all these new Christian elements 'over there' could be found realised and expressed to the same degree among Catholics. This is true even of matters proper to the faith as explicitly developed.[111]

The authentic religious merit of Protestantism meant, therefore, that there was no place for what could be called a 'Catholic triumphalism'. Indeed, Rahner stressed that the convert from

[109] 'On Conversions to the Church', *TI* iii. 375 ('Über Konversionen', *ST* iii. 444), orig. pub. 1953.

[110] Ibid. 376 (ibid.).

[111] Ibid. 377 (ibid. 445).

Protestantism to Catholicism might actually experience a sense of loss:

> Those who are in the religious sense absolute beggars will find everything in the Church when they enter it to be fine and satisfying. But this is precisely what converts are not, if they come from a vital Protestant Christianity to the Church. They will in fact fail to find much in their 'concrete' Church which they are entitled to expect and search for and will in fact lose many things which perhaps they have had up till now. They will comfort themselves with the faithful consciousness that they have done what they had to do to find Christ where he wants to be found, in his Church.[112]

Even while acknowledging the virtues of Protestantism, Rahner saw no contradiction in the Catholic Church's practice of accepting individual converts while continuing to work for Christian unity. Such conversions, he argued, were consistent with the Church's conviction that the various Christian confessions were not simply different expressions of the one reality. So long as the Catholic Church continued to proclaim itself as necessary for salvation, it could not turn away those who sought membership.[113]

When, almost a decade later, Rahner returned to the theme of individual conversions, his ideas had clearly been influenced by his reflections on the shape of contemporary belief—the details of which will be examined in Chapter 4. Thus, while he remained convinced of the value of individual conversion, he was also aware that such conversions were becoming less likely. Consequently, in discussing the reasons why even those who knew and understood the claims of Catholicism could still choose to remain Protestants, the later article introduced the idea that the lack of converts was to be explained not simply by individual influences, but by the general milieu of whole ages and peoples.[114]

Rahner identified two major influences which the social milieu of the late twentieth century had had on ecumenism in general: first, he argued that a modern anti-Christian, neo-pagan spirit, which threatened all Christian denominations, actually made less urgent the question whether one Church was to be preferred to another. In such an environment, even those who did not accept

[112] Ibid. 384 (ibid. 453). [113] Ibid. 375–6 (ibid. 443–4).

[114] 'Some Remarks on the Question of Conversions', *TI* v. 319 ('Einige Bemerkungen über die Frage der Konversionen', *ST* v. 361), orig. pub. 1962.

that denominational differences were irrelevant could acknowledge that having faith was more important than the denominational issue.[115] Secondly, the fact that contemporary developments challenged all the Christian Churches meant that a greater priority ought be given to co-operation between them. Accordingly, Rahner urged the various Christian denominations not only to be open to one another, but also to work together against the common threat of neo-paganism.[116] Such a strategy was to be preferred to a relentless quest to seek converts from each other.

In addition to his reflections on the new social situation of the Churches, Rahner's realization of the changing nature of religious practice in the twentieth century also affected his perspective on the value of individual conversions. Thus, without denying the objective importance of belonging to the Catholic Church, he suggested that since its members often subjectively 'realized' only a small proportion of what the Church had to offer as aids to salvation, those outside the Church who prayed, kept the Commandments, and heeded the word of God could not be regarded as deprived of that grace which the Catholic shared through the sacraments.[117] That this was so, meant that converts of the future would be only those who were likely to pursue more than a merely nominal identification with the Church.[118]

The 1962 article also introduced Rahner's conviction that some of the factors which actually separated the Churches, factors such as their forms of piety and canon law, were historical and human, rather than specifically theological. This did not mean, however, that they were unimportant. Indeed, he contended that the influence of these cultural factors was so pervasive and powerful that it even affected the way specific theological differences were perceived.[119] This theme will be significant when dealing with Rahner's post-Vatican II approach to ecumenism.

REVIEW

The primary implication of Rahner's sacramental ecclesiology is that the Church, because it is the symbol of Christ's presence

[115] 'Some Remarks on the Question of Conversions', *TI* v. 320 ('Einige Bemerkungen über die Frage der Konversionen', *ST* v. 362), orig. pub. 1962.

[116] Ibid. 328–9 (ibid. 371–2).

[117] Ibid. 324–5 (ibid. 367–8).

[118] Ibid. 326 (ibid. 368).

[119] Ibid. 329 (ibid. 372).

in history, becomes central to God's plan for the salvation of humanity. Accordingly, even though Rahner stressed that membership of the Church did not provide a passport to heaven, he was insistent that the offer of grace, the offer of God's own life, was made through the Church. At the same time, his awareness that ever-greater numbers of people had no contact with the Church inspired his efforts to identify a connection—albeit indirect or even 'anonymous'—between those outside the Church and the grace offered through the Church.

Rahner's theory of the 'anonymous Christian' aptly illustrates his commitment to articulating how all aspects of human experience could be understood in relationship to God. As interpreted by some of his critics, however, Rahner's bridging of the gap between those inside and outside the Church was achieved at the expense of the Church itself.

Thus, Medard Kehl questions how it is possible for Rahner to portray the Church as anything more than superfluous, if the grace of salvation is actually offered outside the Church.[120] While Kehl has identified the possibility that Rahner's theology could result in a split between 'the Church' understood as a transcendental reality, and the concrete Church existing in history, it must also be acknowledged that Rahner never denied that even the 'anonymous Christian' was in fact oriented towards the Church and that until the anonymous response found its home in the Church 'something is missing from the fullness of its due nature'.[121]

Kehl's further criticisms—that Rahner's understanding of the Church is unconnected with biblical realities, and that Rahner attempts to deduce the Church from an internal experience of God's self-communication rather than dealing with it as a historical reality—can best be addressed by considering Rahner's theological method.[122] While it is certainly true, particularly in this early period, that Rahner did not establish the scriptural basis for the Church, and that he endeavoured to link the Church to the universal human experience of the supernatural existential and God's self-communication, this does not mean that he was at

[120] Kehl, *Kirche als Institution*, 189–90.

[121] 'Anonymous Christianity and the Missionary Task of the Church', *TI* xii. 164 ('Anonymes Christentum und Missionsauftrag der Kirche', *ST* ix. 501), originally pub. 1970.

[122] Kehl, *Kirche als Institution*, 186–9.

odds with either Scripture or Christian tradition. Indeed, much like his use of Church teaching as an 'end' which was not to be contradicted, but which was also not the final word, Rahner was seeking new categories to express the truths contained in both Scripture and tradition. As part of this endeavour, he attempted to relate Christian experience to an understanding of what it meant to be human.[123] Whether he did so successfully can, of course, be debated; that he sought to do so is, however, fully reconcilable with the vocation of the theologian, especially of a theologian concerned to make Christian truth more comprehensible to his or her contemporaries.

Rahner's sacramental ecclesiology also made a major contribution to his understanding of change in the Church. Thus, the acknowledgement of the sinfulness of the Church—the gap between the symbol and what it symbolizes—established the need for reform in order that the Church might become in fact what it is by nature. Similarly, Rahner's advocacy of ecumenism was also a recognition that change essentially involved the elimination of abuses. In terms of the categories discussed in the Introduction, it will be clear that Rahner's early view of change was very much in keeping with an organic model which equated change with improvement.

Throughout this chapter, the accent has been on Rahner's portrayal of the Church as making present in history Christ's saving grace. What has not yet been discussed, apart from the question of the sinfulness of the Church, is how such a claim relates to the concrete reality of the Church. Does it mean, for example, that the Church's appearance in history, as well as its particular shape, was a miracle, accomplished without the assistance of any human agency? In order to answer such questions, the next chapter will concentrate on Rahner's understanding of the constitution of the Church, both in its divine and human elements.

[123] For a defence of Rahner's method, and its relationship to history, see Leo J. O'Donovan, 'Orthopraxis and Theological Method in Karl Rahner', *Catholic Theological Society of America: Proceedings*, 35 (1980), 47–65.

2

The Shape of the Church

As an enactment of arcane customs, there is little to rival the procedures of the British Parliament. The role of Black Rod, the Lord Chancellor's Woolsack, and the reverence for the mace all contribute to a richly theatrical atmosphere. While Westminster would be the poorer without them, they are none the less all equally tangential to the unromantic realities of late twentieth-century politics. The decline of these once serviceable features of parliamentary life to quaint curiosities provides us with a sobering perspective from which to contemplate the structures of the Church.

In light of the Westminster example, two questions can be asked about the Church's structures: What justification can be given for the existence of the Church's structures, particularly for its hierarchical constitution? and Are these structures capable of responding, without loss of identity, to new situations, or are they, too, destined to become 'museum pieces'? The aim of this chapter is to explore Rahner's response to these questions. The first question can best be addressed via the notion of *ius divinum*; the second, by investigating the dynamics of the connection between the episcopate and the primacy.

To begin, it is important, in the light of Rahner's strictures against an excessive emphasis on the externals of the Church, to acknowledge his basic commitment to the legitimacy of structures in the Church. In fact, the affirmation of a structured Church was an unavoidable corollary of Rahner's adoption of the sacramental model of ecclesiology. In terms of his sacramental theology, structures are important not only because they make the Church a concrete reality in history, but, to apply a concept derived from Rahner's theology of the symbol, because they also symbolize God's presence in history. Without structures, the Church could no longer be the sacrament of the incarnate Word.

This positive endorsement of their role in the Church does not, however, actually convey anything about the specifics of ecclesial structures. It does not address, for example, the question whether it is enough that there simply are structures, or whether these structures must be of a particular kind. Equally, it gives no information concerning the extent to which, or even whether, the Church's structures, even those considered essential to its self-realization, are able to be changed. Such issues are, however, pivotal for an understanding of Rahner's fundamental ecclesiology. What will be evident in discussing them is that, in the period before Vatican II, Rahner's attitude to the structures of the Church was shaped by his acceptance of the notion that they embodied *ius divinum* for the Church.[1] While his application of this idea will be discussed throughout this chapter, his definition of it provides a useful point of entry.

THE CONCEPT OF IUS DIVINUM

As defined by Rahner, in an article first published in 1962, *ius divinum* referred to what was derived from neither natural law nor positive ecclesiastical law. It was, therefore, what could be traced back to 'a positive ordinance of the founder of the Church, an ordinance which—because it comes from him—is by this very fact declared to have been instituted as a permanent feature of the Church and to be an ordinance which even the highest authority of the Church cannot repeal'.[2]

In this category were to be found: papal primacy; the existence of seven sacraments; the need to confess serious sin; the establishment of the episcopate as successors to the apostles; and the threefold hierarchical ministry.[3] Despite the fact that these aspects

[1] For some historical background of *ius divinum*, particularly as understood by Aquinas, Luther, and the Council of Trent, see Avery Dulles, *A Church to Believe in: Discipleship and the Dynamics of Freedom* (New York, 1982), 82–7.

[2] 'Reflections on the Concept of "Ius Divinum" in Catholic Thought' ['Ius Divinum'], *TI* v. 219 ('Über den Begriff "Ius Divinum" im katholischen Verständnis', *ST* v. 249).

[3] For the teaching of Vatican I's *Pastor Aeternus* on papal primacy, see DS 3053–5; for Trent's doctrine, from *De Sacramentis*, on the institution of the seven sacraments by Christ, DS 1601; on the need for confession, from *Doctrina de Sacramento Paenitentiae*, DS 1679; for the episcopal constitution of the Church, from *Doctrina de Sacramento Ordinis*, DS 1768; and, from the same document, the threefold hierarchy, DS 1776—it is worth noting that while the document refers to bishops and priests, it identifies the third level not as deacons, but as *ministris*, which would have included the now-abolished 'minor orders'.

of the Church's life had been defined as *ius divinum*, Rahner recognized that the force of such a claim was less than self-evident. The reluctance to concede that aspects of the Church's constitution were indeed the embodiment of the divine will had two sources: first, the fact that such structures appeared indistinguishable from merely human constructs; secondly, the relationship to history of the structures named above.

To appreciate Rahner's answer to the first objection, it is necessary to return to his theology of the symbol discussed in the previous chapter. In discussing symbols, Rahner defended the claim that even having a social and juridical constitution did not disqualify something from being able to be the symbol of a divine reality. Indeed, he argued that when a free decision was not only to be proclaimed by the symbol, but actually also to be made in it, the juridical composition and free establishment of that symbol were not simply tolerable, but even appropriate.[4] To illustrate this claim, Rahner employed the 'Yes' of the wedding ceremony. In the context of marriage, it was this 'Yes', itself a juridical formula, which was the symbolic reality. That it was not merely an extrinsic sign was proven by the fact that without such a 'Yes', there was no marriage. It could, therefore, legitimately be regarded as the symbol of that bond.[5]

Since the Church too existed as a social entity, since it too had to be freely chosen, the same principle could be applied to it. Consequently, the Church's juridical constitution could not, a priori, be regarded as contradicting its claim to be the symbolic reality of Christ's redemptive presence.[6] This was especially so because the Church's existence as the sacrament of Christ, the Spirit-filled creation of Christ's redemptive act, meant that it was never simply a juridical-social organization.

Rahner acknowledged, however, that the fact that each of those aspects of the Church's structure for which a divine initiative was claimed could in fact be shown to have evolved seemed to gainsay any divine origin. This was especially a difficulty when the original forms of such institutions bore little resemblance to their later shape.[7] Such scepticism was further strengthened by the awareness that, at the time of their first appearance, these early forms

[4] 'Symbol', *TI* iv. 240 (*ST* iv. 297).

[5] Ibid. (ibid. 298). The discussion on the marriage vows can be found in n. 13 in both the Eng. and Ger. texts.

[6] Ibid. 241 (ibid.). [7] 'Ius Divinum', *TI* v. 221 (*ST* v. 250).

had not been regarded as either unchangeable or valid for all time.[8] The challenge for Rahner was, therefore, to show how this history of development could be reconciled with the belief that the structures involved actually fulfilled a divine ordinance. Examination of his response to this difficulty can best begin with what alone could furnish the possibility that particular aspects of the Church's life were indeed the direct expression of God's will: revelation.

REVELATION: GOD'S SELF-COMMUNICATION

The distinguishing feature of revelation theology in this century has been its lack of consensus. Indeed, in his survey of the field, Avery Dulles lists five major models—each of which has also fathered sundry permutations—which have been utilized in an effort to expound revelation.[9] Rahner too contributed to this proliferation.

The two elements characteristic of Rahner's treatment of revelation were: his rejection of the notion that revelation could be understood as God handing on to humanity a set of propositions about God; and his championing of the idea that the essence of revelation was a dialogue between God and humanity, the 'hearer of the word'. These characteristics can be clearly seen in a 1954 article which discussed revelation in the context of the development of dogma:

> Revelation is not the communication of a definite number of propositions, a numerical sum, to which additions may conceivably be made at will or which can suddenly and arbitrarily be limited, but an historical dialogue between God and humanity in which something *happens*, and in which the communication is related to the continuous 'happening' and enterprise of God. This dialogue moves to a quite definitive term, in which the *happening* and *consequently* the communication comes to its never to be surpassed climax and so its conclusion. Revelation is a saving Happening, and only then, and in relation to this, a communication of 'truths'.[10]

[8] 'Ius Divinum', *TI* v. 221 (*ST* v. 251).

[9] Avery Dulles, *Models of Revelation* (New York, 1983). His analysis of the various models can be found on pp. 36–128.

[10] 'The Development of Dogma' ['Development'], *TI* i. 48 ('Zur Frage der Dogmenentwicklung', *ST* i. 59). Rahner's emphasis.

The 'saving Happening' initiated by God was God's self-communication to humanity. In order to set the scene for specifying the status of ecclesial institutions and doctrines, what needs to be investigated here are the dynamics of the human reception of God's self-disclosure. In other words, the focus will be on Rahner's claim that the human person has a *potentia obedientialis* for the revelation of God.[11]

In the previous chapter, in discussing Rahner's approach to the relationship between grace and nature, the point was made that, by means of the supernatural existential, Rahner identified openness to God as a constitutive element of human existence. While the supernatural existential provided the basis for a connection between God and humanity, the importance of that concept cannot be fully grasped without reference to Rahner's epistemology.

The key aspect of Rahner's epistemology in relation to revelation was that revelation, although it always remained God's free act, must take place in a way consistent with the contours of human knowing. If this was not done, human beings would never be able to hear the word spoken by God. This principle had two major specifications: first, because human beings exist in history, God's self-communication had to take place in history. Revelation was not, therefore, merely an internal mystical experience for which human existence in history was utterly irrelevant.[12] Secondly, a corollary of the above, since human beings came to knowledge by means of sense perception—*nihil sine phantasmate intellegit anima* —God's self-revelation could not be independent of concrete human experiences.[13] Once these epistemological values were set in place, Rahner combined them with metaphysics to produce his distinctive theology of revelation.

While Rahner endorsed the principle that human knowing could never bypass objects of the senses—the categorial—he was also insistent that our knowing was not restricted to sense objects.

[11] *Hearers of the Word* [*Hearers*], rev. edn. Johann B. Metz, trans. M. Richards (Montreal, 1969), 67 (*Hörer des Wortes: Zur Grundlegung einer Religionsphilosophie* [Hörer], 2nd edn. (Freiburg i.B., 1971), 77). The orig. Ger. edn. was pub. in 1941, but in 1963 Johann Baptist Metz—with Rahner's approval—revised the work. Both the Ger. and Eng. refs. in this book are to the Metz version. While there is no complete Eng. trans. available of the first Ger. edn., large sections of it, trans. by Joseph Donceel, can be found in Gerald A. McCool (ed.), *A Rahner Reader* (New York, 1981), 2–65.

[12] *Hearers*, 156–9 (*Hörer*, 167–71).

[13] Ibid. 146–9 (ibid. 157–60).

Thus, he argued that in knowing any object, we not only recognized it as limited, as not the cause of all possible objects, but also had a sense of something 'more'—the transcendental.[14] This 'more' was not merely another object—if it were, we would also have a sense of something beyond it—but 'the opening up of the absolute breadth of possible objectivity in general'.[15] Our recognition of the limitedness of every particular object was based on a 'pre-grasp' or *Vorgriff* of a 'horizon' which was not itself an object but 'is being and thus the totality of all possible objects of human knowledge'.[16] The *Vorgriff*, therefore, which was part of every act of cognition, manifested our orientation not simply to other beings, but to absolute being (*das Sein*).[17] In other words, the 'whither' (*Woraufhin*) of the *Vorgriff* was towards absolute being.[18] Our 'pre-grasp' affirmed a being beyond the realm of the imagination. The next step for Rahner was to identify this absolute being with God.

Rahner argued that while the *Vorgriff* did not present God as an immediate object of the intellect, the operation of the *Vorgriff* affirmed the existence of God as the 'existent thing of absolute "having being"' ('das Seiende absoluter "Seinshabe"').[19] In other words, through the affirmation of the finitude of existent things, the affirmation of the existence of an *esse absolutum* also took place.[20]

To complete the foundations for his theology of revelation, Rahner combined metaphysics and epistemology with the anthropology which underpinned the supernatural existential:

A revelation from God is thus possible only if the subject to whom it is supposed to be addressed presents *in him- or herself* an a priori horizon

[14] *Hearers*, 59 (*Hörer*, 68–9). [15] Ibid. 59 (ibid. 69).

[16] Ibid. 59–61 (ibid. 69–71). For the background and development of the *Vorgriff*, particularly in regard to its relationship to Thomas, see: Paul J. de Rosa, 'Karl Rahner's Concept of "Vorgriff": An Examination of its Philosophical Background' (Univ. of Oxford Ph.D. thesis, 1988); Jack Arthur Bonsor, *Rahner, Heidegger, and Truth: Karl Rahner's Notion of Christian Truth—The Influence of Martin Heidegger* (Lanham, Md., 1987), 75–85; see also George Vass, *Understanding Karl Rahner*, i: *A Theologian in Search of a Philosophy* (London, 1985), 31–64; and, Otto Muck, *The Transcendental Method*, trans. W. D. Seidensticker (New York, 1968), 184–204.

[17] *Hearers*, 60–1 (*Hörer*, 70–1).

[18] *Spirit in the World*, trans. W. Dych (New York, 1968), 398 (*Geist in Welt*, 2nd edn. (Munich, 1957), 397). For Rahner's early ideas on the *Vorgriff*, before its fuller development in *Hearers*, see pp. 142–5 (pp. 155–8). This book, orig. pub. 1939, was Rahner's doctoral thesis, which was rejected by his examiners in Freiburg i.B. For details of the that rejection, see *I Remember*, 41–4 (*Erinnerungen*, 42–6). The edn. cited here was ed. by Metz and pub. in 1957.

[19] *Hearers*, 63–4 (*Hörer*, 73–4).

against which such a possible revelation can begin to present itself in the first place. . . . A revelation which is supposed to disclose the depths of divinity, and which at bottom is the reflex objectification of the human calling to participate in nothing less than the supernatural life of God, can only be conceived as possible if the human being is conceived as spirit.[21]

Humanity's God-given orientation to God, when combined with the fact that God was encountered—even if not named—as the 'horizon' of all categorial experiences, meant that God's self-communication to humanity was not a rare occurrence, but was the distinguishing feature of human existence. Even though every created object spoke of God, there was none the less one presence of God in human history which was more than a mere horizon:

This positive nature of creation . . . reaches its qualitatively unique climax in Christ. For, according to the testimony of the faith, this created human nature is the indispensable and permanent gateway through which everything created must pass if it is to find the perfection of its eternal validity before God. . . . Whoever sees Christ, sees the Father, and whoever does not see him—God become man—also does not see God. We may speak about the *impersonal* Absolute without the non-absolute flesh of the Son, but the *personal* Absolute can be truly found only in him, in whom dwells the fullness of the Godhead in the earthly vessel of his humanity.[22]

Rahner interpreted the title 'saviour' to mean that, in Christ, God's self-communication became irrevocable. Since any historical movement was to be understood in terms of its end, the whole event of God's self-communication—even as it had taken place before Christ—could be grasped in Christ:

The whole movement of this history lives only for the moment of arrival at its goal and climax—it lives for its entry into the event which makes it irreversible—in short, it lives for the one whom we call Saviour. This Saviour who represents the climax of this self-communication, must therefore be at the same time God's absolute pledge by self-communication to

[20] Ibid. 64 (ibid. 74). Not surprisingly, Rahner's approach to human knowledge of God has not been without its critics; see e.g. Joseph O'Leary, *Questioning Back: The Overcoming of Metaphysics in Christian Tradition* (Minneapolis, 1985), esp. 87–98, and Paul Molnar, 'Can We Know God Directly? Rahner's Solution from Experience', *TS* 46 (1985), 228–61.

[21] *Hearers*, 66–7 (*Hörer*, 76–7). Rahner's emphasis.

[22] 'Humanity of Jesus', *TI* iii. 43 (*ST* iii. 56–7). Rahner's emphasis.

the spiritual creature as a whole *and* the acceptance of this self-communication by the Saviour; only then is there an utterly irrevocable self-communication on both sides, and only thus is it present in the world in a historically communicative manner.[23]

Prior to Christ, it was a matter for conjecture whether salvation or damnation would actually be God's definitive answer to the fickleness of humanity. Only in Christ were such doubts resolved.[24] Only in Christ—'the indissoluble, irrevocable presence of God in the world as salvation, love and forgiveness, as communication to the world of the most intimate depths of the divine reality itself and of its Trinitarian life'—did God display an unambiguous commitment to creation.[25] Consequently, Christ was central to any theology of revelation.

After, and because of, Christ, there could be neither a new age of human history nor a different plan for salvation: there could be only the 'unveiling' of what had been definitively communicated in him.[26] After Christ, nothing more could be said, since, in him, everything possible had been said. For this reason, even the idea of the closure of revelation could be understood positively. This closure indicated that what had already been communicated to humanity was nothing partial, but rather the 'fulfilled presence of an all-embracing plenitude'.[27] Revelation was closed because it was complete. After Christ, human experience of God was to be interpreted in the light of Christ.

For his efforts to demonstrate that a proven history of development was not necessarily incompatible with something existing *iure divino*, what is of primary importance in Rahner's view of revelation is the claim that even God's definitive revelation in Christ needs to be 'unveiled'. This assertion raises two questions: Why is such an unveiling necessary? and What are the dynamics of the process of unveiling? The first question will be answered in what follows; the dynamics of the process, which involve the work of the Spirit in the Church, will be explored in the next chapter.

[23] 'Christology within an Evolutionary View of the World', *TI* v. 175–6 ('Die Christologie innerhalb einer evolutiven Weltanschauung', *ST* v. 203). Rahner's emphasis. Orig. pub. 1962. Critical assessments of Rahner's Christology can be found in Bruce Marshall, *Christology in Conflict: The Identity of a Saviour in Rahner and Barth* (Oxford, 1987), 15–114, and John M. McDermott, 'The Christologies of Karl Rahner', *Gregorianum*, 67 (1986), 87–123, 297–327.

[24] 'Development', *TI* i. 49 (*ST* i. 59). [25] Ibid. (ibid. 60).

[26] Ibid. (ibid.). [27] Ibid. (ibid.).

In analysing why Rahner considered an unfolding of revelation to be necessary, the focus will primarily be on his understanding of the relationship between revelation and dogma. Lest this seem too major a digression from the theme of *ius divinum* and the Church's structures, it is worth noting the parallels between the two topics: both involve the conversion of the human experience of God into forms consistent with the concrete, historical nature of human existence; both involve the idea of development; and both illustrate the nexus between revelation and the Church, especially the essential role of the apostolic Church and the Church's teaching authority. For these reasons, an analysis of how dogmatic statements arise out of the experience of revelation can serve as an introduction to an exploration of the Church's structures.

REVELATION: THE EXPRESSION OF ITS CONTENT

A good example of Rahner's view of the need for the unveiling of revelation is found in his 1962 article, 'Theology in the New Testament', the basic thesis of which is that an unveiling of revelation takes place even within the New Testament itself. In that article, in order to illustrate that revelation was more than a one-dimensional process, Rahner drew an analogy between revelation and mystical experience. In his analysis of mysticism, he stressed that there was a need to distinguish between the mystic's initial awareness of God and the subsequent reflex objectification, interpretation, and conceptual communication of that encounter.[28] The event itself was both primary and not exhausted by efforts to expound it. The richness of the mystical experience was such that no human statements could reproduce it. None the less, such statements were indispensable, as without them communication of the experience was not possible.[29]

From this foundation, Rahner claimed that a similar distinction could be drawn between the original event of revelation and its objectification in Scripture and dogmatic formulations.[30] Once

[28] 'Theology in the New Testament', *TI* v. 40 ('Theologie im Neuen Testament', *ST* v. 52).

[29] Ibid. 40–1 (ibid. 52–3).

[30] Ibid. (ibid.). For a discussion of Rahner's approach to the relationship between revelation and dogma, see Mary E. Hines, *The Transformation of Dogma: An Introduction to Karl Rahner on Doctrine* (New York, 1989). 31–49.

revelation was understood as an 'event' or 'happening', as God's self-communication rather than the reception of information about God, it was clear that revelation too could never be definitively expressed by a set of propositions. The effort to elaborate the implications of God's self-communication, to unveil the revelation, was, however, intrinsic to the human response to God: the experience and the words belonged together. While the experience of revelation was not to be equated with those statements which the Church proclaimed as revelatory, it was, nevertheless, the basis of what was objectified in them.

The human need to objectify experiences in order to communicate them meant that words were an indispensable element of authentic communication. Such a conclusion accords with Rahner's emphasis on the fact that God's self-communication to humanity took place within history, and via appearances, not in a non-historical rationalism. In other words, God's revelation was attuned to the dynamics of human existence—'the revelation of God, whatever its origin, must in the end be translated into human speech, if human beings are not to be completely taken out of their human mode of existence by it'.[31]

Although theological statements were thus a necessary aspect of the reception of God's self-communication, they were none the less inadequate to the task of expressing the fullness of God. This inadequacy derived from the limits imposed by human languages. Since dogmatic statements had to be brief, comprehensible, and accommodated to the faith consciousness of a broad audience, the full reality of God's self-communication was not to be found in them.[32] Although such statements could highlight certain aspects of God's infinite fullness, they inevitably left others shrouded in darkness.[33] That human statements about God failed to elucidate the whole of the divine reality was not, therefore, the result of poor formulation, but of their incommensurability to their task. Just as a description of being in love is inferior to the actual experience of loving, so Rahner argued that no propositions could exhaust the richness of God's self-communication.[34] While

[31] *Hearers*, 158 (*Hörer*, 169–70). See also Bonsor, *Rahner, Heidegger, and Truth*, 141–53.

[32] 'What is a Dogmatic Statement?' ['Dogmatic Statement'], *TI* v. 54 ('Was ist eine dogmatische Aussage?', *ST* v. 68), orig. pub. 1961.

[33] Ibid. 55 (ibid. 69).

[34] 'Development', *TI* i. 64–6 (*ST* i. 76–8).

dogmatic statements could point the way to God, they could not encapsulate everything which could be known about God. God was always greater:

The dogmatic statement—like the kerygmatic one—is basically possessed of an element which (in the case of intra-mundane categorical statements) is not identical with the represented conceptual content. Without injuring its own meaning, the represented conceptual content is in this case merely the means of experiencing a being referred beyond itself and everything imaginable. That this reference is no mere empty, frustrated desire to transcend, that it is not simply the formal horizon for the possibility of objective conceptualisation, but the way in which human beings really move towards the self-communication of God as God is in God's self is brought about by what we call grace and accepted in what we call faith.[35]

Rahner stressed, however, that characterizing human statements about God as 'limited' was not a euphemism for 'false'.[36] What their limited nature did imply was that, while retaining their truth, they could be surpassed by statements which were both more nuanced and which opened up realities and truths not grasped by previous formulations.[37] In addition, since each doctrinal statement was also limited, as will be seen shortly, by the fact that it necessarily reflected the perceptions and terminology appropriate to the period in which it was formulated, new conditions in social and intellectual history could make necessary new formulations.[38] Having identified the reasons why Rahner believed that changes in doctrinal formulations might become necessary, the next task is to explore his analysis of the processes involved in forming new perceptions of revealed truths, the process of the development of dogma.

THE DEVELOPMENT OF DOGMA

While Rahner sought to provide a rational analysis of how development occurred, he did not attempt to produce a priori laws of development. On the contrary, he argued that what was possible in terms of development could only be understood from analysis

[35] 'Dogmatic Statement', *TI* v. 58–9 (*ST* v. 73).
[36] 'Development', *TI* i. 44 (*ST* i. 54). [37] Ibid. (ibid.).
[38] 'Dogmatic Statement', *TI* v. 55 (*ST* v. 69).

of what had actually taken place.[39] In 'Considerations of the Development of Dogma', the second of two articles which he wrote on this theme in the 1950s, Rahner claimed that since development involved the activity of human beings, it could not be made the subject of rigid laws. Although it was thus impossible to frame universal laws of development, he was convinced that development would certainly take place:

If humanity has a history, not only as a physical, a biological living thing, but even as spirit, indeed, if it is firstly as spirit that humanity possesses a true history, it must be clear from the beginning that this history takes in fact a unique course and is not the constant repetition of the same law. This must be true above all of the sublimest portion of the history of the human spirit, the history of the divine revelation in the human spirit and the unfolding of this revelation. It would be remarkable if there was a history of divine revelation—which no Christian can deny—but no history of the unfolding of this revelation. . . . There is then no adequate formal theory of the development of dogma which would be in itself sufficient to permit a prognosis for the future.[40]

Such a claim could be justified by the fact that, as was indicated above, the inability of dogmatic statements to capture divine transcendence in its fullness meant that they were always imperfect and, therefore, open to further refinement. For this reason, reference to the closure of revelation was not to be understood as meaning that humanity had nothing left to learn about God. Far from implying that humanity had mastered God, the closure of revelation was best understood as describing 'humanity's being opened up for and into the real and not merely conceptual self-communication of God. And therein it has within itself precisely because of this closure, which is disclosure, its dynamism of inner development, and hence a dynamism of the development of dogma.'[41]

If it was God's transcendence which made development inevitable, it was the fact that the human person was both spiritual and existed in history which provided the catalyst for development to take place. Historical existence, which meant that each age interpreted its awareness of God in a unique way, suggested that no

[39] 'Development', *TI* i. 41 (*ST* i. 50).

[40] 'Considerations on the Development of Dogma' ['Considerations'], *TI* iv. 8 ('Überlegungen zur Dogmenentwicklung', *ST* iv. 17), orig. pub. 1958.

[41] Ibid. 9 (ibid. 19).

one formulation of belief could meet the needs of every period of history.[42] Spirituality, as a result of which human beings, unlike a photographic plate, did not indifferently register each new experience, provided the impetus for bringing any new insights into relationship with what had been previously perceived.[43] Both the spirituality of human beings and their existence in history suggested, therefore, that development did not mean the abandonment of what already existed. Just as new theories in philosophy did not imply that there was nothing left to learn from Plato, so too the development of dogma built on Scripture, the Fathers, and the scholastics without making any of them redundant.[44]

Consequently, Rahner argued that change was not to be equated with progress, with the Church becoming more clever, but with a different way of perceiving the divine reality, a way appropriate to humanity at a particular stage of its history. Development was, therefore, a change in, not of, identity ('der Wandel im selben').[45] The challenge to Christians was not that of choosing between development and preservation, but of recognizing that both were essential. To deny the possibility of development was to fail to appropriate the truth in a new setting; to repudiate preservation was to risk error by ignoring accumulated wisdom.[46] Preserving the balance between the claims of change and continuity was the vocation of the Church.

DEVELOPMENT AND THE CHURCH

In order to appreciate the pivotal role of the Church in the process of development, it is necessary to amplify the previous chapter's discussion of the relationship between God and the Church. In his 1958 study of the inspiration of Scripture, Rahner contended that the Church was part of God's design for the Incarnation of the Logos, that it existed by a 'pre-definition' in the context of salvation history.[47] It was, therefore, valid to distinguish between God's willing of the Church and God's willing of the rest of creation; to assert that, without obliterating human freedom, God had specifically appropriated the Church as a part of the plan for

[42] 'Development', *TI* i. 45 (*ST* i. 55).
[43] Ibid. 44 (ibid. 54). [44] Ibid. 45 (ibid. 56).
[45] Ibid. (ibid.). [46] Ibid. (ibid.).
[47] *Inspiration*, 40 (*Schriftinspiration*, 47).

the salvation of the world in Christ.[48] Through Christ, the Church was the sign of the new covenant, the sacrament of God's irrevocable commitment to humanity.

Since the Church alone had the whole history of salvation for the experience and development of its faith, Rahner argued that its faith was always more comprehensive than that of the individual believer.[49] As a result, the individual was both humbled and liberated by believing with the Church: humbled, as faith lived within the Church implied a recognition of the limits of one's personal faith; liberated, because through the faith of the Church all believers shared a breadth of wisdom and understanding which transcended the insights of individuals.[50] Thus, it was the belief of the Church which was the standard of belief. Indeed, Rahner argued that all theology ought be addressed, either implicitly or explicitly, to the whole Church, as only a theology accepted by the whole Church could be regarded as an authentic expression of Christian truth.[51]

Its singular role in the history of salvation also afforded the Church a unique competence to determine what constituted an authentic unfolding of revelation. Thus, it was the Church, since it presided over the unfolding of revealed truth, which enabled believers to take possession of that truth in each new historical context.[52] In order to do this, the Church had to ensure that continuity and development remained complementary.

Just as the processes of development could not be reduced to exceptionless norms, so too the balance between development and continuity could be negotiated only in a historical context which was itself evolving. This context could, however, endanger the prospects of either continuity or development, or both. Thus, writing in the late 1950s, Rahner acknowledged that commitment to continuity meant resisting the attacks of historicism, liberal Protestantism, and Modernism.[53] Equally, the efforts of Catholic apologists to maintain that development was no more than minor changes in words, the warning in *Humani Generis* against any historical relativizing of Church dogmas, and the assaults of Protestant neo-orthodoxy, which, on the basis of a strict interpretation

[48] *Inspiration*, 41–2 (*Schriftinspiration*, 47–8).
[49] 'Ecclesiological Piety', *TI* v. 346 (*ST* v. 389).
[50] Ibid. 347 (ibid. 390–1).
[51] 'Dogmatic Statement', *TI* v. 53 (*ST* v. 67).
[52] 'Considerations', *TI* iv. 3 (*ST* iv. 11).
[53] Ibid. 5 (ibid. 13).

of *sola scriptura*, regarded Catholic dogmatics as prone to novelty, made it difficult to promote the legitimacy of development.[54] The responsibility for effecting the marriage of continuity and development belonged, in the name of the whole Church, to the Church's teaching authority.

In *The Dynamic Element in the Church*, which he wrote in 1958, Rahner stressed that Christ's promise to remain with his Church applied in a particular way to the Spirit's guidance of the teaching authority.[55] The specifics of this claim, as well as its implications for the relationship between office-bearers and the rest of the Church, will be examined in the next chapter, which focuses on the Spirit in the Church. For the present, it is sufficient to recognize that the necessity of hearing the word from the authoritative teaching office was intrinsic to Rahner's view of development:

> The word of God is always the word delivered by the authorized bearer of doctrine and tradition in the hierarchically-constituted Church. The element of being officially-delivered, the reference back to an authorized teacher, the hearing of the word *from* this authoritative teaching person, belongs to the constitutive moments of dogma and hence also of dogmatic development. The development of dogma never takes place without the teaching authority. Dogma must be proclaimed as such by the magisterium, the teaching authority, in order that it can be believed *fide ecclesiastica* with the Church, as a constitutive moment of the faith of this *Church* itself.[56]

Rahner also acknowledged that the ecclesial character of dogmatic statements meant that even their terminology had to be understood as the Church's terminology. Consequently, dogmatic statements were to be interpreted in the light of the teaching authority's own use of this terminology.[57] As was illustrated in the previous chapter, this was the process Rahner himself followed in beginning with the sense of 'Church' utilized in *Mystici Corporis*.[58] This principle notwithstanding, Rahner also believed—as was

[54] Ibid. (ibid.).

[55] *The Dynamic Element in the Church* [*Dynamic Element*], trans. W. J. O'Hara (London, 1964), 42–3 (*Das Dynamische in der Kirche* [*Das Dynamische*] (Freiburg i.B., 1958), 38–9).

[56] 'Considerations', *TI* iv. 15 (*ST* iv. 25). Rahner's emphasis.

[57] 'Dogmatic Statement', *TI* v. 54–5 (*ST* v. 68–9).

[58] A further application of this same principle can be found in his analysis of *Humani Generis* in, 'Theological Reflections on Monogenism', *TI* i. 229–96 ('Theologisches zum Monogenismus', *ST* i. 253–322), orig. pub. 1954.

clearly indicated by his development of the idea of 'Church' beyond the usage of the encyclical—that this terminology itself needed to develop. Unless this was done, the danger was that the expressions proper to only one period of history would not only become frozen, but would themselves be equated with the reality they were intended to describe.[59]

While harmonizing continuity and development in each new period of history was a complex task, those who exercised the teaching authority of the Church did not act solely on the basis of extraordinary interventions by the Spirit. Rather, they sought to remain faithful to the direction of the path laid by their predecessors, the *depositum fidei*. Rahner argued that the fact that belief had a history was itself a protection against inauthentic development, since the clearer a truth became, the less likelihood there was of future error.[60] This could be illustrated from the natural realm, where, for all living things, every advance had a definitive quality which made regression impossible. Within this process of gradually clarifying the truth, a particular role had been played by the primitive Church. Its contribution was not merely the first stage in the long process of clarifying the truth, but was normative for all other periods of the Church's history.

DEVELOPMENT AND APOSTOLICITY

While there is always a temptation to canonize the past as a golden age, Rahner's claim that the endowment from the apostolic period was normative for the later Church was not based on any such exaggerated respect for antiquity. In fact, it was underpinned by the conviction that development meant a change in, not of, identity. For this reason, the earliest formulations of Christian belief, and the structures developed by the first generation of the Church, necessarily remained the Church's corner-stones, the foundations on which everything else was built.

As was noted when discussing his distinction between the experience of revelation and its subsequent expression, Rahner was convinced that what was received in God's self-communication was always richer than the efforts to capture it in words. In the

[59] 'Dogmatic Statement', *TI* v. 56–7 (*ST* v. 70–1).
[60] 'Development', *TI* i. 42 (*ST* i. 52).

same way, the experience of the first generation of Christians was simpler and more concentrated than the panoply of articulated theology which built on it.[61] The first phase of Christian history enjoyed 'an originality, an irreducibility and a purity in the expression of its own essence which, necessary as subsequent evolution must be, are proper only to the first phase'.[62]

Consequently, not only did this early period of the Church's history exericise a critical role in the articulation of revelation, but its formulations acted as a norm against which later claimants to the authentic unfolding of revelation were to be judged. For these reasons, it is important to be clear on what Rahner understood by 'the early Church' or 'the apostolic Church'. In fact, he generally preferred to speak of 'the primitive Church' (*die Urkirche*) rather than 'the apostolic age'. This was so for two reasons: first, he argued that as the Gospels of Luke and Mark, as well as the Letter to the Hebrews, showed signs of having sources outside the apostles, they could not be considered 'apostolic' if that term was understood too specifically;[63] secondly, the corollary of the above, insistence on the traditional notion that revelation was closed with the death of the last apostle raised problems for those works which were canonical, but which either could not be connected directly with an apostle or could not be shown to have achieved their final form before the death of the last apostle.[64] Where Rahner himself did indeed refer to 'the apostolic Church', he also expressed the desire for this period to be understood in a fluid sense, rather than being tied to a calendar day.[65]

His desire for a more inclusive definition of the earliest period of the Church did not imply, however, that Rahner was seeking to devalue its importance. Indeed, he argued that if the whole history of the Church was under the auspices of the Spirit, then the primitive Church was uniquely so. He illustrated this assertion by an analogy with human experience: what occurred in the primitive Church differed from what came later in the same way that the act of being born and the subsequent act of living do not have the same qualitative connection with the mother.[66] This first age of the Church was to be considered part of God's particular providence for the community of faith:

[61] Ibid. 67 (ibid. 79). [62] *Inspiration*, 46 (*Schriftinspiration*, 52).
[63] 'Ius Divinum', *TI* v. 232 (*ST* v. 263). [64] Ibid. (ibid. 264).
[65] *Inspiration*, 72 (*Schriftinspiration*, 76). [66] Ibid. 46 (ibid. 52).

Thus it would not be correct to say that God's founding of the Church consists simply in conserving it in existence; rather, we must say that an essential part of God's conserving it in existence consists in God having founded it at a particular moment in time. God, then, as founder of the Church, has a unique, qualitatively not transmissible relationship to the Church's first generation, one which God does not have in the same sense to other periods of the Church's history, or rather has to the latter only through the former.[67]

What the first generation of the Church set in place was 'the authoritative "beginning", the permanent norm, the all-supporting foundation and the unsurpassable law, for all subsequent Christianity'.[68] In short, it had a *kairos* which remained. The fact that neither the beliefs nor the structures of the Church were reconstituted in each new period of history meant, therefore, that what came to be later was necessarily dependent on what was already in place. These subsequent developments did not come directly from Christ himself, but from what already existed—'they derive from their foundations, not merely from him who laid the foundations'.[69] The normative role of the apostolic Church made it a matter of urgency to know where the foundations it lay were to be encountered. The two candidates for providing this source were Scripture and tradition.

Far from seeking to resuscitate the controversy generated by regarding Scripture and tradition as two independent sources of revelation, Rahner was anxious to highlight their unity. In order to do so, he needed to connect tradition to revelation. In 'Scripture and Tradition', an article written in 1963, Rahner argued that if revelation involved Christ handing himself over to us—*traditio Jesu Christi*—in his reality, deeds, and words, then such revelation was dependent on an oral tradition which could witness to Christ.[70] This meant that the witness of the primitive Church, those who had directly experienced Christ, was normative. What this Church passed on was not simply a collection of teachings, but what it itself believed and expressed in the sacraments it celebrated.[71]

[67] *Inspiration*, 45 (*Schriftinspiration*, 51).

[68] 'Ius Divinum', *TI* v. 232 (*ST* v. 264). For Rahner's ideas on how it was possible for one period of history to have meaning for later periods, see *Hearers*, 158–60 (*Hörer*, 169–71).

[69] *Inspiration*, 45 (*Schriftinspiration*, 51).

[70] 'Scripture and Tradition', *TI* vi. 101 ('Heilige Schrift und Tradition', *ST* vi. 125), orig. pub. 1963.

[71] Ibid. 102 (ibid. 126).

None the less, the faith of the primitive Church was principally to be grasped in the Scriptures. It was in these Scriptures that the primitive Church interpreted its own experience for later ages. Accordingly, Scripture itself was to be understood as a tradition, as the Church carrying on its witness to the saving event of Christ.[72] Rahner emphasized the Catholic conviction that the apostles and the authors of the New Testament were not just the first generation of transmitters of revelation, but were themselves 'vessels of revelation'.[73] As a result, what they taught was not simply their private theology, but actually embodied revelation. Based on the belief that Scripture was the definitive witness of the primitive Church, Rahner rejected the notion that truths of faith could be known independently of Scripture. Indeed, he argued that later developments in the Church's faith could only be considered authentic if they could be related to what had been preached from the beginning as an explicit element of revelation; in other words, authenticity depended on a relationship to what was to be found in Scripture.[74]

Consequently, Rahner's claim that the teaching authority alone could articulate the faith of the Church is not to be interpreted as elevating that authority above Scripture. Since Scripture, understood as expressing the faith of the primitive Church, was the *norma normans* of the Church's faith, the mission of the teaching authority had to be understood in relationship, not opposition, to it.[75] Indeed, Rahner argued that just as the infallibility of the teaching authority of the apostolic Church lay in its capacity to determine the Canon of Scripture, the infallibility of the later magisterium was to be found in its ability to interpret Scripture inerrantly.[76]

Having examined the principles which Rahner identified as crucial to the development of dogma, the next step is to apply these same principles to the structures existing in the Church *iure divino*. As will be seen, the contentious point is again that of the legitimacy of development: how is it possible for aspects of the Church's constitution to be *ius divinum* if they can be shown to have developed?

[72] Ibid. 102–3 (ibid.). [73] 'Ius Divinum', *TI* v. 231 (*ST* v. 262–3).

[74] 'Virginitas in Partu', *TI* iv. 147 ('Virginitas in Partu', *ST* iv. 187), orig. pub. 1960.

[75] 'Dogmatic Statement', *TI* v. 53 (*ST* v. 67).

[76] *Inspiration*, 77 (*Schriftinspiration*, 81).

IUS DIVINUM AND THE CHURCH'S STRUCTURES

At the beginning of this chapter, the evolution of the Church's structures was identified as one of the two basic obstacles to according them *ius divinum* status. This historical maturation presents a difficulty because it can be interpreted as indicative of human, rather than divine, origin.[77] In the light of this objection, it might be expected that Rahner's priority would be to link such structures to Christ. This was not, however, the case. Indeed, without repudiating his emphasis on the foundational role of the apostolic Church, Rahner rejected the view that only what could be shown to be directly dependent on a statement of Jesus or the apostles could be *ius divinum* and, therefore, immutable.[78]

He opposed such a claim because it would have committed the Church to maintaining traditions which, like the process for fraternal correction in Matt. 18: 15–18, although genuine expressions of apostolic practice, were simply the by-products of a particular culture rather than intrinsic to the work of the Church.[79] Indeed, far from being bound by the mere existence of apostolic ties, Rahner was even prepared, as late as 1962, to contemplate the abolition of the diaconate. Significantly, his ground for so doing was the fact that the diaconate, despite its longevity and its being included in the hierarchy of ministries, which itself existed *iure divino*, no longer served a real function in the Church.[80]

If even a specific connection to Christ did not suffice, how then could the *ius divinum* status of the Church's structures and practices be established? Rahner's answer was that any irreversible, law-establishing decision which was both in conformity with the Church's nature and which took place in the time of the primitive Church was *ius divinum* for the Church.[81] Such a principle not only opened the way for aspects of Church life not associated with a specific expression of Christ to be regarded as *ius divinum*, it also allowed the possibility that even what had not from its first

[77] See Dulles, *A Church to Believe in*, 85–7.

[78] 'Ius Divinum', *TI* v. 225 (*ST* v. 255). [79] Ibid. (ibid.).

[80] Ibid. (ibid. 256). In the same year as he produced this negative view of the diaconate, Rahner also outlined a proposal for its 'rehabilitation': see 'The Theology of the Restoration of the Diaconate', *TI* v. 268–314 ('Die Theologie der Erneuerung des Diakonates', *ST* v. 303–55).

[81] 'Ius Divinum, *TI* v. 230–1 (*ST* v. 262). For further discussion of Rahner's approach to *ius divinum*, see Walther Altmann, *Der Begriff der Tradition bei Karl Rahner* (Berne, 1974), 196–202.

appearance been considered an expression of the divine will could none the less exist *iure divino*. Since the cogency of this claim is not necessarily self-evident, the three arguments which Rahner advanced to support it need to be examined.

First, drawing on philosophy, Rahner argued that as a being or notion in metaphysics could be understood only via a *conversio ad phantasmata*, so *ius divinum* was inseparable from its incarnation in particular historical forms.[82] Furthermore, if the divine will was indeed expressed in concrete, historical realities, it implied that the human decisions which set in place those particular aspects of the Church's life were themselves agents in the process of revelation. Thus, for example, the inspiration of the Gospels could not be understood without regarding the human decision to write them as itself part of the revelatory process. Neither could the connection of the primacy to Rome, nor the revelation of Matthias as God's choice to take a place among 'the Twelve', be so regarded without, respectively, Peter's decision to go to Rome and the actual election of Matthias (Acts 1: 26).[83] The events themselves were not only revelatory, but actually brought about the revelation:

> Events can certainly have the character of a definite material revelation; this presupposes merely that they take place, as it were, in the field of those who already have a certain knowledge of revelation and thus in the light of such knowledge are in a position to judge such an event, and the determined quality of the event which can be known only through revelation. It is not then possible to say that only the material principles are revealed and not the quality of the event which is known with the help of these principles. At least one will not be able to say this as far as the apostolic age is concerned. . . . Free events, therefore, can certainly have the character of revelation in the primitive Church. They are human decisions, and in them precisely is accomplished God's will to reveal, a will which desires, brings about and reveals this event in all its characteristic nature in and through this freedom of decision pre-defined by this will.[84]

Secondly, while acknowledging that the shape of what was introduced into the Church by such decisions could indeed change over a given period, Rahner argued that to interpret such changes

[82] 'Ius Divinum', *TI* v. 221–2 (*ST* v. 251–2).
[83] Ibid. 233–4 (ibid. 265–6). [84] Ibid. 234–5 (ibid. 266–7).

as producing a stream of new realities was to confuse essence and form.[85] His counter-assertion, in line with his claim that the development of dogma involved a change in, not of, identity, was that the historical development of Church structures illustrated that it was possible for the same nature to persist under different forms.[86]

Thirdly, he claimed that the later emergence of features identified as *ius divinum* could not be accepted as proving comprehensively that they were either unnecessary or reversible. This view was in response to the argument that, since what the Church accepted as *ius divinum* was, by definition, essential to its constitution, those features of its life so identified ought to have been present at its foundation or to have been initially accompanied by the declaration that they were indeed the embodiment of irreversible divine law. In rejecting this assertion, Rahner argued that while it was true that whatever belonged to the nature of a spiritual, personal reality in the physical-social order must be present in some way as soon as the thing itself was present, this did not mean that its full flowering had to be perceptible when the reality was in its embryonic state.[87]

In the light of the above, Rahner was prepared to accept as *ius divinum*, and, therefore, as irreversible, the law-establishing decisions taken in the period of the primitive Church. Such decisions could be revelatory even if they involved choosing between various options which had been equally possible at the time. What was crucial was not that the particular decisions themselves could be shown to have been necessary, but that they could be shown to be in conformity with the nature of the Church.[88] In addition, such decisions could be, and could have been regarded at the time, irreversible and binding on the Church for all ages:

> the revealed character of such a *ius divinum* does not exclude the observable nature of its development from empirically verifiable tendencies and causes which were in a kind of competitive combat with other existing tendencies of development. Even the divine law of the Church is a divine-human law. Even the life of the law, and hence also of the divine-human law, is a one-way history in which (similar to the phylogenesis and ontogenesis of the living being) the concrete form of the law develops out

[85] 'Ius Divinum', *TI* v. 222 (*ST* v. 252). [86] Ibid. (ibid.).
[87] Ibid. 230 (ibid. 261). [88] Ibid. 239 (ibid. 271).

of a necessarily multi-potential system by progressively determining what is to be realised out of the great abundance of the potential.[89]

To complete this examination of Rahner's approach to *ius divinum*, what remains to be asked is whether it is possible that something which emerged after the apostolic age could still be regarded as an expression of the divine will for the Church. On the one hand, this question would seem to be pointless because, as has already been seen, Rahner stressed the indispensable mediation of the primitive Church. On the other, however, since he recognized that not everything which the Church accepts as *ius divinum* could be shown to have been in place definitively by the end of the apostolic period, he also warned against too hastily denying the possibility of post-apostolic *ius divinum*.[90]

Although he was unusually coy about revealing whether he did or did not accept post-apostolic *ius divinum*, Rahner still raised some provocative questions. His stalking-horse was the status of the decisions of the Church's magisterium. Rahner's basic thrust was that if it was accepted that the magisterium's decisions in the post-apostolic period could be infallible, then there was a need to account for this infallibility. As outlined by Rahner, there were two possible foundations for this infallibility: either such decisions could be regarded as revelation, that is the word of God, to be accepted in 'divine faith'; or they could be classified simply as the word of the Church, to be accepted in 'ecclesiastical faith'.[91] While the latter demanded respect because of the divinely grounded authority of the Church, it could not be regarded as revelation, nor, presumably, as infallible. If, however, the notion of 'ecclesiastical faith' was rejected as having no basis in tradition, then the sole remaining option was to believe that revelation occurred even in the post-apostolic Church.

Whether Rahner himself held this opinion in the pre-Vatican II period is, in contemporary diplomatic jargon, impossible to confirm or deny. The most that can be gleaned from his writings is the somewhat sibylline declaration that the feasibility of post-apostolic revelation, and therefore *ius divinum*, ought not to be 'too quickly and unreflectingly' rejected.[92]

In surveying Rahner's defence of *ius divinum*, the point was

[89] Ibid. (ibid. 272). [90] Ibid. 241 (ibid. 274).
[91] Ibid. 242 (ibid. 275–6). [92] Ibid. 243 (ibid. 277).

made that he believed the notion of an irreversible divine law for the Church could properly be understood only when incarnated in particular historical forms. The final task of this chapter, therefore, is to analyse his pre-Vatican II approach to two such expressions of the Church's constitution: the priesthood, and the relationship between the episcopate and papal primacy. In examining the former, what will be emphasized is the necessity of interpreting the priesthood in relationship to the Church, not as an end in itself; in regard to the latter, the focus will be on answering the question posed at the beginning of this chapter—are those structures regarded as existing *iure divino* able to develop?

THE PRIESTHOOD: ESSENCE AND MISSION

In an article on the apostolate of the laity, which he wrote in 1955, Rahner distinguished the laity from the clergy on the grounds that only the latter exercised 'the apostolate of an official mission'. In addition, he claimed that this mission, since it 'annulled' the priest's original position in the world and involved 'founding a new existence', needed the grace of its own sacrament.[93] The aim of this section is to explain how Rahner understood this 'new existence' of the priest.

Prior to doing so, however, it should quickly be pointed out, lest the above taint Rahner with today's unpardonable sin of 'clericalism', that he derived the difference between clergy and laity from their respective functions within the Church, rather than from personal qualities. Indeed, he specifically rejected any implication that the personal holiness of the laity, or the extent to which lay people were embraced by the love of God, was inferior to that of the clergy.[94] In addition, he not only accepted that all Christians had a role to play in making Christ present in the world—indeed, as early as 1936, he had spoken of all the baptized as consecrated, through baptism, to the care of souls[95]—but claimed

[93] 'Notes on the Lay Apostolate' ['Apostolate'], *TI* ii. 335 ('Über das Laienapostolat', *ST* ii. 356).

[94] See e.g. 'The Sacramental Basis for the Role of the Layman in the Church' ['Layman'], *TI* viii. 72 ('Sakramentale Grundlegung des Laienstandes in der Kirche', *ST* vii. 349), orig. pub. 1960.

[95] 'The Consecration of the Layman to the Care of Souls' ['Care of Souls'], *TI* iii. 274 ('Weihe des Laien zur Seelsorge', *ST* iii. 325).

that a particular duty of office-bearers in the Church was to help the laity in their witness to Christ.[96] As a result of this view, Rahner even argued that it was possible to maintain that the clergy existed only because the laity existed.[97] If the clergy were not necessarily holier than other members of the Church, how then was their 'new existence' to be understood? This question can best be answered by following Rahner's inquiry into the notion of priestly 'character' in his 1942 article, 'Priestly Existence'.

Crucial to Rahner's presentation of the priesthood was relating it to the Church's dependence on Christ. He stressed, therefore, that priests were not to be understood as 'mediators' where this suggested that they bridged a gap between God and God's people. Priests were, rather, the sacramental signs of what had been achieved by Christ, and by him alone.[98] As a result of this unique mediation by Christ, the priest could not claim originality in either his cultic or prophetic ministry, the two areas of activity fundamental to the priestly life.

In regard to its cultic activity, the Church's priesthood was only a 'ministerial priesthood'. This was so as it derived not from the priest himself, but from the priest- and victimhood of Christ.[99] In addition, the Eucharist, which was the central feature of the priest's cultic activity, belonged to the whole Church, not just its ministers. Rahner did emphasize, however, that the manner of celebrating the Eucharist expressed the hierarchical structure of the Church. He regarded it, therefore, as 'according to the will of Christ' that a priest alone could preside at the Eucharist.[100] Nevertheless, the fullness of Christian life which the priest, like all Christians, obtained from the Eucharist was always the result of his receiving, rather than distributing, the sacrament.[101] While affirming the uniqueness of the priest's cultic function, Rahner regarded this function as insufficiently 'existentially foundational' to be the basis of a specifically priestly character. Cultic activity was excluded from being central to priestly character not only because it did not occupy the whole life of the priest, but also because its efficacy was independent of the quality of the priest's efforts.[102]

Of greater existential significance was the prophetic role of the

[96] 'Layman', *TI* viii. 67 (*ST* vii. 345). [97] Ibid. (ibid.).
[98] 'Priestly Existence', *TI* iii. 248 (*ST* iii. 295–6).
[99] Ibid. 249. (ibid. 296). [100] 'Layman', *TI* viii. 69–70 (*ST* vii. 347).
[101] Ibid. 70 (ibid.). [102] 'Priestly Existence', TI iii. 256–7 (*ST* iii. 305).

priest. Here there were no restrictions to the commitment demanded of the minister. Since the preaching of the Gospel was more than a lecture, it required the total personal engagement of the preacher; without this involvement, his preaching would not necessarily be false, but could simply be the imparting of religious knowledge rather than the call to faith.[103] Rahner also asserted that the manner in which the priest gave witness to faith was to be distinguished from that of any other Christian as the priest alone was commissioned to witness not simply to his own faith, but to the authentic word of Christ:

The official priestly herald of the message of Jesus utters his word as one who alone is empowered to make present cultically the salvation reality of Christ. And because of this, as well as by the mission which lies upon him from Christ through the apostolic succession, he utters his word not as one addressed by the Word of God; he does not bear witness to his *own* Christian existence as such—even though this is for him and his hearers an indispensable condition of a proper proclamation—but he speaks the word of Christ itself as such.[104]

Although he acknowledged that the witness of the non-ordained Christian could often be more effective than that of the clergy, Rahner nevertheless highlighted a difference in scope between the two. While the vocation of the lay Christian was to witness to Christ in their immediate situation, the clergy alone were charged to provide this witness always and everywhere.[105] Since the priest was always the one sent 'from above', it was this commission, not his own situation in the world, which empowered the priest to be involved even where others would see him as interfering in what they regarded as their private affairs.[106] It was because this task of the priest engaged the whole person that it involved a new existential significance, a unique character.

There are several aspects of Rahner's early interpretation of the priesthood which will be significant in later sections of this book: first, the fact that he portrayed the priest as primarily a minister of the Church, the sacrament of Christ, suggested the need for a developing theology of the priesthood as the Church's self-understanding developed; secondly, because Rahner began his

[103] 'Priestly Existence', *TI* iii. 259 (*ST* iii. 308).
[104] Ibid. 260 (ibid. 309–10). Rahner's emphasis.
[105] Ibid. 261 (ibid. 310). [106] Ibid. (ibid. 310–11).

analysis of the priesthood with baptism, rather than with the usual scriptural references, he could be understood as implying that the priesthood ought be open to all the baptized; thirdly, his rejection of cultic activity as constitutive of priestly character opened the way for an understanding of priestly ministry beyond that which had been dominant in the centuries since the Council of Trent; fourthly, while Rahner had articulated a theology linking the priest to the mission of the Church, his clear distinction between the priest and the laity—the laity shared in the official priestly ministry only to the extent that they supplied the means of life to those who lived 'from the altar'[107]—meant that the role of the non-ordained in the mission of the Church remained to be addressed.

Another aspect of the Church's hierarchical constitution which Rahner explored in the years before Vatican II was the relationship between the bishops and the Pope as primate of the universal Church. What is significant in Rahner's treatment of this theme is that, in his major study, *The Episcopate and the Primacy*, which he co-authored with Joseph Ratzinger in 1961, he accepted that this relationship too existed in the Church *iure divino*, but none the less argued that its dynamics could develop far beyond the model which had prevailed since Vatican I. His understanding of how the papacy and the episcopacy were connected, as well as his proposals for the evolution of this bond, will be discussed in what follows.

THE EPISCOPATE AND THE PRIMACY

In outlining how *ius divinum* applied to the Church's structures, Rahner emphasized that it was possible to distinguish between the essence of what was God's will for the Church and the particular form in which that essence was realized, a form which could itself develop.[108] His treatment of the connection between the episcopacy and the primacy illustrates the meaning of this principle. Thus, while accepting the legitimacy of papal primacy, he also sought to protect the authentic rights of the episcopate against any tendencies to centralism. Crucial to his approach was asserting the integrity of the local Church. To aid in this crusade, Rahner employed the Old Testament notions of 'the remnant' and the synagogue.

[107] Ibid. (ibid. 310). [108] 'Ius Divinum', *TI* v. 223–4 (*ST* v. 253–4).

These he interpreted as indicating that the local Church, including its bishop, could not be viewed simply as an agent of the universal Church.

What was significant for Rahner in the concept of 'the remnant', was the idea that the whole could be present in the part.[109] In the context of contemporary ecclesiology, this meant that 'the Church' could be claimed as present in a local community: 'the Church as a whole, where it becomes "Event" in the full sense, is also necessarily a local Church, the whole Church becomes tangible in the local Church'.[110]

Not merely the notion of the remnant, but also the history of the synagogue suggested the need for greater recognition of the local Church. This was so because the individual synagogue had been regarded not simply as a fragment of the whole nation, but rather as itself the manifestation of the whole.[111] Consistent with this idea, Rahner claimed, in a 1956 article on the parish, that each local Church—understood in this context as the parish rather than the diocese—enabled the Church as a whole to realize its 'historical, place-time apprehensibility'.[112]

In order to substantiate this emphasis on the local Church, the role of the Eucharist, where the Church was 'event' *par excellence*, was decisive. This was so for two reasons: in being the activity of the eschatological community of salvation, the celebration of the Eucharist revealed both the presence of Christ and the unity of believers, with Christ and one another; since all this happened in the local Church, it also made clear that this Church was truly 'the Church'.[113] Without prejudice to the universal nature of the Church, therefore, such an interpretation of the Eucharist highlighted that the local Church was not merely an agent of the universal Church, but was itself an event of that Church.

One consequence of such an interpretation of the local Church was that the bishops, as chief pastors of the individual communities,

[109] *The Episcopate and the Primacy* [*Episcopate*], trans. K. Barker, P. Kerans, R. Ochs, and R. Strachan (New York, 1962), 23 (*Episkopat und Primat* [*Episkopat*] (Freiburg i.B., 1961), 23).

[110] Ibid. (ibid. 24).

[111] 'Theology of the Parish', in Hugo Rahner (ed.), *The Parish: From Theology to Practice*, trans. R. Kress (Westminster, Md., 1958), 27 ('Zur Theologie der Pfarre', in id. (ed.), *Die Pfarre: Von der Theologie zur Praxis* (Freiburg i.B., 1956), 31).

[112] Ibid. 26 (ibid. 30). [113] *Episcopate*, 25–6 (*Episkopat*, 26).

were to be regarded as embodying the Church at the local level. Although the Pope alone represented, by divine right, the unity of the Church as the totality of local Churches, Rahner stressed that the Pope's power of order was no more comprehensive than that of a local bishop.[114] This implied, for example, that the local bishop was not obliged simply to relay the teaching of the Pope, but could himself authoritatively witness to revealed truth. While the Pope had full, immediate, ordinary and general episcopal primacy of jurisdiction over the whole Church and each of its members, individual bishops included, this did not mean that he was an absolute monarch. An absolute monarchy in the Church was in fact inconceivable, since the episcopate as a whole, because it also existed *iure divino*, could not be disciplined or abolished by the Pope.[115] Although the bishops were indeed appointed by the Pope, their episcopal authority did not derive from his personal authority, but from what belonged, *iure divino*, to the episcopate itself.[116]

The task of harmonizing the rights of the papacy and the episcopate for the well-being of the whole Church belonged in a particular way, as will be seen in the next chapter, to the Spirit. Nevertheless, Rahner believed there was a need for greater precision in delineating the rights peculiar to the episcopate. If this was not done, then it could appear that, apart from the formal protection against abolition, all the rights of the episcopate were derived, *iure humano*, from the Pope.[117] The key to providing the necessary clarity lay in recognizing that the individual bishops and the Pope himself were all members of the college of bishops.

Rahner regarded the college of bishops as the successor of the apostolic college founded by Christ. The members of this college, he argued, derived their power neither directly from Christ himself nor from Peter, but from their membership of the college.[118] The primacy of Peter reflected his headship of the college, but did not imply that the college was founded by him:

> Ontologically and juridically, then, the apostolic college, with Peter as its head, forms a unity. It does not exist without Peter, nor Peter without it. One could say: Peter is appointed by Christ as head of the Church

[114] Ibid. 29 (ibid. 29).
[115] Ibid. 16–17 (ibid. 17–18).
[116] Ibid. 70 (ibid. 66).
[117] Ibid. 69 (ibid. 64).
[118] Ibid. 77–8 (ibid. 72–3).

insofar as he is head of the apostolic college which he rules, while ruling the Church *with* it.[119]

What appealed to Rahner in the idea of the college was its making concrete of the fact that the Church was not simply a unity, but a unity of parts.[120] A pluralism which was not a threat to the unity of the Church was, therefore, not simply to be tolerated, but to be preserved and promoted. The college of bishops ensured that this pluralism, which was of divine right in the Church, was also seen in its head.[121] Only by the bishops being dispersed in their local Churches throughout the world, rather than being grouped together in Rome—where they would have appeared to be the Pope's personal presbyterate—could this necessary pluralism, and the *iure divino* prerogatives of the episcopate, be realized:

The bishops are 'local ordinaries' not merely because the Pope for practical reasons needs administrative officials in various places for his personal rule of the Church . . . but because a bishop can fulfil the function which he has *in* the college of bishops and indeed for the *entire* Church, only when he authoritatively represents a particular member of the universal Church, in which the differentiation from other members of the Church intended by the Spirit can really exist.[122]

Building on this foundation, Rahner characterized as heresy any suggestion that there could be only one form of liturgy or law, or that the homogeneity of Church life in the West—which, as will be discussed in a later chapter, he actually regarded as a historical accident—ought to be the norm for the whole Church.[123] He argued that as nations drew closer together even without the need for a single language, so the Church ought to open itself more to varied local expressions rather than reserving all decisions, great and small, to Rome.[124] The presence of the Spirit, the existing principles of unity in law and dogma, and the example of an accepted pluralism in the Churches of the Eastern Catholic rites all indicated that pluralism and unity were not contradictory. If regional developments were supported by the local bishop, this

[119] *Episcopate*, 79 (*Episkopat*, 74). Rahner's emphasis.
[120] Ibid. 105–6 (ibid. 97–8). [121] Ibid. 107 (ibid. 99–100).
[122] Ibid. 108. (ibid. 100–1). The emphasis in the text, which does not appear in the pub. Eng. trans., is Rahner's.
[123] Ibid. 106 (ibid. 98). [124] Ibid. 113–14 (ibid. 105–6).

provided the guarantee that they were not damaging to the Church's unity.

As a right, not a gift, each bishop was, therefore, to be left room for personal initiative. Only by thus preserving its own character, could his diocese make a contribution to the whole Church.[125] A bishop acting in this way was not being presumptuous, but simply fulfilling his duty. This duty was also fulfilled even when the acts of the local Church could not be used as a model for the whole Church, but simply met a local need. The arena available for local initiatives was indeed broad. It included liturgy, the life-style of the faithful, catechetics, and religious life.[126] In addition, Rahner also championed the view that ecumenism could take particular local forms that might not be able to be applied simultaneously to the whole Church.[127]

The desire to promote both the local Church and authentic pluralism also led Rahner to encourage the development of conferences of bishops on a national or continental level. In an article written in 1963, he argued that such conferences, although they were a late development, could nevertheless be regarded as 'an absolutely necessary expression of an essential element of the Church'.[128] Indeed, Rahner was even prepared to advocate that larger administrative units of the Church, and not simply the individual bishop in his individual diocese, could actually put into practice *iure divino* aspects of the Church's constitution.[129] He stressed that, since the permanent nature of the Church had to be made present in historical situations which were themselves ever-changing, it was always valid to question whether particular aspects of the Church's constitution could be brought out more clearly or lived more effectively.[130] His support for bishops' conferences was an application of his commitment to such development.

Although accepting that the Spirit was the source of the proper relationship between the episcopate and the primacy, Rahner was also not averse to making use of legislation to clarify certain issues. His particular concern was to eliminate ambiguities which

[125] Ibid 110–11 (ibid. 103). [126] Ibid. 112 (ibid. 104).

[127] Ibid. 116–17 (ibid. 108–9).

[128] 'On Bishops' Conferences', *TI* vi. 377 ('Über Bischofskonferenzen', *ST* vi. 441).

[129] Ibid. 378 (ibid. 442).

[130] 'The Episcopal Office', *TI* vi. 320 ('Über den Episkopat', *ST* vi. 377), orig. pub. 1963.

allowed Roman congregations to treat local bishops as if they were provincial officials of a central government.[131] Thus, he was anxious to give legal status to bishops' conferences and to specify the procedures for a Council so that it could not be commandeered by curial officials—a particularly pertinent issue on the eve of Vatican II.[132]

REVIEW

At the beginning of this chapter, it was asked whether the Church's structures could be considered both necessary to its existence in history and capable of development. Rahner's answer to these questions was a resounding 'Yes'. As has been noted, he not only believed that certain ecclesial structures could be interpreted as an expression of God's will for the Church, but he also insisted that their permanence was not inimical to a capacity for development. Accordingly, as was apparent in his analysis of the relationship between the episcopate and the primacy, he did not regard even those institutions whose permanence could be ascribed to *ius divinum* as being beyond the influence of history, human law, and practical experience. In addition, as was shown by his stress on Scripture and the apostolic Church as norms for legitimating later developments, Rahner's commitment to change existed alongside his defence of continuity within the Church.

Rahner's defence of the legitimate rights of the local bishop against the encroachments of a centralizing authority, and his emphasis on the Church becoming 'event' in the context of the local celebration of the Eucharist—an emphasis which significantly pre-dated *Lumen Gentium*—both demonstrated that acknowledging a divine law in the Church need not be synonymous with portraying the Church as a petrified monolith. Significantly, his early advocacy of the priority to be given to the local Church appears to have been shaped by theological rather than historical considerations. As will be indicated, however, his awareness of changes in the Church's position in the world became increasingly important not only for his promotion of the centrality of the local Church, but also for his understanding of the possible shape of priestly ministry.

[131] *Episcopate*, 118 (*Episkopat*, 109–10).
[132] Ibid. 120 (Ibid. 111).

Perhaps not surprisingly, Rahner's interpretation of the Church's structures has not been without its critics. One point of attack has been the lack of specific references to Scripture in his analysis of the origins of those structures. Thus, Ursula Schnell sees in Rahner's approach to Church structures a reliance on social theory—the need for any group to organize and institutionalize itself if it is to survive beyond the first generation—rather than God's movement in history. This criticism derives both from Rahner's lack of biblical references to the origins of the Church's structures and from his emphasis on the Church as freeing believers from the dangers of individualism.[133] Similarly, Medard Kehl argues that Rahner's approach to office and the social organization of the Church would have been more convincing if his ecclesiology had relied more on 'concrete, biblical content' rather than the 'formal abstractness' of his approach to sacramentality.[134]

Rahner's use—or, more correctly, non-use—of Scripture is defended by Robert Kress on the grounds that to insist on a biblical reference for each of the Church's structures is to presume an abyss between humanity and grace. Indeed, Kress accuses Schnell of a 'biblical positivism and theological abyssism' for failing to appreciate that the natural need of human beings for orderliness is not on a different plane from the development of an ordered Church.[135] Certainly, it would not accord with Rahner's understanding of the graced nature of humanity to regard only those activities buttressed by a direct scriptural imperative as legitimately Christian. Consequently, his discussion of the formation of the Church's structures in the apostolic era ought not be regarded as implying that this formation was merely a human activity. In addition, neither Schnell nor Kehl acknowledges Rahner's recognition, discussed above, that only what is reconcilable with the revelation given in Scripture could be regarded as divine will for the Church. Indeed, as was claimed in the previous chapter, Rahner's method represents an interpretation, not a rejection, of the witness of Scripture.

[133] Ursula Schnell, *Das Verhältnis von Amt und Gemeinde im neueren Katholizismus* (Berlin, 1977), 210–13. Schnell's criticisms are actually applied to Rahner's post-Vatican II work, but, if correct, they would seem far more relevant to his earlier writings.

[134] Kehl, *Kirche als Institution*, 221.

[135] Robert Kress, *The Church: Communion, Sacrament, Communication* (New York, 1985), 170–1.

In the light of the foregoing criticism that his approach to the Church's structures is mere sociology, it is ironic that Rahner is also accused of not being sociological enough. Thus, Leo Dullaart finds Rahner's work lacking in a theory of, or even a proper appreciation for, institutions. Dullaart alleges that Rahner merely tolerates the Church's institutional nature, and that he continually subordinates it to the primacy of individual salvation.[136]

While it can be freely admitted that Rahner did not develop a comprehensive theory of institutions, the force of Dullaart's criticism is none the less problematic. Rahner was not concerned with institutions *qua* institutions but, as will be seen, with the concrete activity of the particular institutions which existed in the Church. Significantly, Dullaart's criticism of Rahner does not acknowledge any of Rahner's assessments of the actual performance of the Church's structures in history. In addition, Dullaart does not discuss the essential role which structures have in Rahner's sacramental approach to the Church—indeed, there is no consideration of Rahner's notion of sacramentality. Finally, Dullaart's concern that Rahner ignores the 'self-worth' of institutions in favour of their role in the salvation of the individual seems to imply the danger of isolating institutions from any meaning beyond themselves.[137] As has been emphasized, however, while Rahner did indeed prize the institutional aspects of the Church, he was not willing to regard institutions as an end in themselves.

Criticism of a different order comes from Avery Dulles, who is sceptical about the legitimacy of Rahner's interpretation of post-apostolic *ius divinum*. According to Dulles:

> Once one admits that *ius divinum* may depend upon a development in time, it is difficult to insist upon absolute irreversibility. What is appropriate or even necessary for a later age is admitted to have been inappropriate or even impossible for an earlier time. If this is so, how can we say that in some future time or in some other culture the previous development might not again become inappropriate or impossible? If development is acknowledged, the institution which develops becomes tied to certain historical and cultural conditions whose permanence might seem to be questionable. Thus the theory of development seems to call for something like de-development, at least as an abstract possibility.[138]

[136] Leo Dullaart, *Kirche und Ekklesiologie* (Munich, 1975), 148–53.

[137] Ibid. 150–1.

[138] Dulles, *A Church To Believe in*, 91–2.

The first point that must be made in answering Dulles is to insist that unless *ius divinum* is actually linked to history, the whole concept can appear as something magical—an arbitrary intervention by God—and unrelated to the concrete situation of the Church. In addition, to accept the historical development of *ius divinum* is to acknowledge that revelation unfolds in history. Such a view is consistent with Rahner's emphasis that the explication of revelation involves being drawn more deeply into the divine mystery, rather than mastering a limited body of truths. Furthermore, Dulles's fear that historical development is potentially incompatible with irreversibility can be answered from Rahner's notion, which will be developed in a later chapter, that continuity involves an act of faith, rather than the mere maintenance of what was present from the first moments of the Church's history. Rahner's position would mean, therefore, that the Church had a greater freedom than is usually acknowledged to develop even its *iure divino* structures.

Before considering in detail Rahner's assessment of the Church's response to history, what remains to be explored is the final aspect of his theological analysis of the process of change: the activity of the Holy Spirit, the agent of change in the Church. Accordingly, the next chapter will focus on the charismatic dimension of Rahner's ecclesiology.

3

The Spirit: 'The Element of Dynamic Unrest'

ONLY after the Ceausescu dictatorship was overthrown in 1989 did the extent of the 'Big Brother' activities of the Securidade, that regime's secret police, become apparent. What was revealed was a system which was both cruelly oppressive and irredeemably Kafkaesque. A prime illustration of its absurdity was the network of informers operated by the police: the regime's obsession with invigilating the population's every thought and action meant that there were so many informers—perhaps one person in six—that it proved impossible to process the information they generated. Consequently, most police files, which recorded every minute detail of life in Romania, were never read. In that system without a soul, it became impossible to question the value of such an operation, impossible to do otherwise than perpetuate the absurdity. This loss of reason within the system itself was compounded by its brutal humiliation of those who dared to cry out that the emperor was indeed naked.

While the internal workings of a dictatorship are an extreme case, they none the less graphically illustrate the dangers that can result when a system becomes an end in itself, when it loses the capacity to question its own priorities and processes, and when those outside the ruling élite are not permitted to call attention to defects in the system. With these dangers in mind, the present chapter will examine Rahner's understanding of both the animation and adaptability of the Church. In short, the chapter will be concerned with the Holy Spirit in the Church.

In Rahner's theology, the Spirit and the Church were inextricably linked. If its existence as the sacrament of Christ in history meant that the Church needed a visible form, it also meant that the Church needed a source of animation: this animation was provided by the Spirit.

Rahner's exposition of the link between the Spirit and the Church can be clearly seen, for example, in a 1956 article, 'The Church as the Subject of the Sending of the Spirit':

> In Jesus of Nazareth we have the living God of the living Spirit and of grace. The Church is nothing else than the further projection of the historicity and visibility of Jesus through space and time and every word of its message, every one of its sacramental signs is, once more, nothing else than a part of the world in its earthiness with which the Spirit has united itself indissolubly since the day on which the Logos became flesh.[1]

Making his own the conviction of Irenaeus that where the Church is, there too is the Spirit of God, and vice versa, Rahner argued that there could be no talk of the Spirit without reference to the Church.[2] What lacked this bond to the Church was, *ipso facto*, not an expression of the Spirit, but merely religious enthusiasm. The very existence of the Church revealed that humanity remained embraced by the life of Pentecost. Thus, following the lead of Pius XII in *Mystici Corporis*, Rahner asserted that the role of the Spirit in the Church was significant enough to justify the belief that the Church had a double structure: not only was it hierarchical, it was also charismatic.[3]

Rahner's commitment to the relationship between the Spirit and the Church was also a commitment to change in the Church. Indeed, as early as 1946, he characterized the mission of the Spirit in the Church as that of being the 'element of dynamic unrest—if not of revolutionary upheaval'.[4] That he maintained his view throughout the period prior to Vatican II is evident in 'Do Not Stifle the Spirit!', an article which appeared shortly before the opening of the Council. In this work, he emphasized that it was the Spirit who could move the Church in directions which were not part of the Church's planning.[5] Explaining why the Spirit should have such a role will be part of what follows.

[1] 'The Church as the Subject of the Sending of the Spirit' ['Spirit'], *TI* vii. 188–9 ('Die Kirche als Ort der Geistsendung', *ST* vii. 185).

[2] Ibid. 189 (ibid. 186).

[3] 'Saints', *TI* iii. 103–4 (*ST* iii. 124–5). The ref. in *Mystici Corporis* to the charismatic structure of the Church can be found in DS 3801, 3807–8.

[4] 'The Individual in the Church' ('Individual'), in *Nature and Grace* [*Nature*], trans. D. Wharton (London, 1963), 79 ('Der Einzelne in der Kirche', in *Gefahren im heutigen Katholizismus* [*Gefahren*] (Einsiedeln, 1955), 35).

[5] 'Do Not Stifle the Spirit!' ('Stifle'), *TI* vii. 74–5 ('Löscht den Geist nicht aus!', *ST* vii. 79).

Rahner's interpretation of the Spirit's activity in the Church can best be explored via three principal themes, each of which has two, potentially divisive, elements: the Spirit as the source of both the Church's truth and of the development of its doctrine; the particular gifts of the Spirit given both to those inside and outside of the Church's hierarchy; and the Spirit's role as the origin of both individuality and community in the Church. As part of each of these topics, Rahner's emphasis on the Spirit as the source both of unity and of diversity in the Church will be highlighted.

THE DYNAMICS OF THE CHURCH'S GROWTH IN FAITH

As depicted by Rahner, the mission of the Spirit was neither simply to preserve the Church from error, nor to remodel it in every generation. Instead, the Spirit was the heart of that process which, as was discussed in the last chapter, was integral to Rahner's view of revelation: the 'development and unfolding of the original treasure of faith'.[6] The role of the Spirit in that process of unfolding will be discussed in the next section; in the present section, the focus will be on the human processes involved in the deepening of the Church's understanding of revelation.

For Rahner, the fact that revelation was God's self-communication, rather than of a set of formulas about God, ultimately determined the processes of unfolding revelation. If revelation had indeed involved nothing but a string of propositions, logical deduction would have sufficed as the principle of explication. Such a method could not, however, do justice to the richer experience of God's self-revelation. To appreciate why this should be so, there is a need to venture back into the jungles of Rahnerian epistemology. This can be done with minimal hardship by following his differentiation between mathematical propositions—the proper subjects of logical deduction—and personal ones—the means by which we, and God, reveal ourselves.

While the content of a mathematical proposition could be strictly delineated, so that no more was communicated through such a formula than could be deduced from an analysis of its component parts, human statements were more complex. This complexity did not imply, however, that human statements were unintelligible.

[6] 'Development', *TI* i. 52 (*ST* i. 63).

What it did suggest was that the clear meanings which could be established for human propositions were always only a minimum, which, in contrast to the situation in mathematics, did not exhaust the range of legitimate possible interpretations of such statements.[7] Whereas mathematical axioms functioned like a package with univocal, clearly defined contents, human propositions were more like a window that opened on to wider vistas:

> For example, if I say, 'N.N. is my mother', what has been communicated by this? What have I thought of here and *communicated* [*mitgeteilt*]? The minimum is clear: what would make the proposition false if it did not exist: such and such a well-known biological relation. But do we mean by this that the proposition has nothing more to communicate, that when I spoke I thought of nothing more and had nothing more to state? When I make a statement like this, there can and almost must be an abundance of other things in mind at the same time. . . . But *all this*, in excess of the given minimum of propositional content, can concomitantly be heard by the hearer of the proposition. . . . The hearer too, just like the speaker, looks in and through the proposition with the speaker at the thing itself, and sees what he sees in the thing *as* the communication made to one who is taking part as hearer.[8]

Thus, according to Rahner's analysis, human communication always involved more than the hearer simply receiving what the speaker positively articulated. Propositions, therefore, could not constitute the sum total of communication. Indeed, Rahner asserted that the propositions which expressed the content of any act of communication could conceivably be formulated by the hearer, as a result of reflecting on what had been communicated, rather than by the speaker.[9]

In the context of revelation understood as God's self-disclosure, Rahner's claim that communication involved more than the contents of reflex propositions suggested that the process of development could not be satisfactorily explained by portraying the Church as simply deducing what was formally implicit in the earlier expressions. In fact, Rahner regarded such a procedure as merely a hermeneutical or exegetical operation, which, although it resulted in new forms of expression, did not produce any authentically new knowledge.[10] The alternative view, the one favoured by Rahner,

[7] Ibid. 68–9 (ibid. 81–2).
[8] Ibid. 69–70 (ibid. 82). Rahner's emphasis.
[9] Ibid. 70–1 (ibid. 83–4).
[10] Ibid. 59 (ibid. 71).

was that development involved articulating those levels of meaning which were only virtually, rather than formally, implicit in earlier expressions. Such levels of meaning, while not included in the actual formulations used, were none the less part of what was communicated with the explicit statements of faith; as such, they could not be retrieved solely by the machinery of syllogistic logic.[11] Development entailed, therefore, the articulation of a genuinely new awareness. While this awareness was certainly not independent of what had been expressed in previous formulations, it was also more than simply an exegesis of their terms.

As regards the Church's unfolding of revelation, Rahner's approach suggested that development occurred when the Church used existing formulations of doctrine as a 'window' to come to a deeper insight into the meaning of God's act of self-disclosure, the revelation which those earlier dogmatic formulations had also sought to articulate. It was the Church, the sacrament of Christ, which, as was discussed in the previous chapter, was uniquely capable of recognizing what was new, what was an authentic unfolding of revelation rather than a mere theologumenon.[12] Without implying that it was an addition to God's definitive self-communication, this process could correctly be called revelation, since, in deepening its awareness of God, the Church was actually attaining new knowledge.[13]

While the intricacies of interpersonal communication are best reflected not by the mathematical model, but by Rahner's contention that such communication always involves more than is immediately stated, it ought none the less to be asked whether his approach actually provides a satisfactory paradigm for God's self-communication. Is there a danger, for example, that Rahner's analysis implies that God, no less than human beings when we speak, in fact communicates more than God is either aware of or intends? If so, it would suggest that the Church is able to find in God's self-communication meanings which God had in fact not intended; meanings which, therefore, could not be identified genuinely as God's self-revelation. Rahner's answer to this conundrum was to claim that all the possible consequences of God's act of self-communication were both known and intended by God:

[11] 'Development', *TI* i. 72–4 (*ST* i. 85–7).
[12] Ibid. 74–5 (ibid. 88–9). [13] Ibid. (ibid.).

God is necessarily conscious of the actual vitality and dynamism of God's immediate communication, and aware of all its virtues and consequences. Moreover God has from the very beginning the intention and the will to bring about its explication and to guide it in God's own Spirit. . . . Virtual explication, looked at from the point of view of *God* as speaker, is simply explication and nothing more, even if it requires real deduction by *us*, looked at from our point of view as hearers. What *we* 'deduce' in this way, God has not indeed *stated* 'formally' in the initial propositions from which our deduction proceeds (that is, God has not expressed it in the immediate meaning of the propositions), but God has really 'communicated' [mit-geteilt] it, so that entire faith can be given to it as *God's* knowledge.[14]

Although stressing that the Church deepened its knowledge of God's self-communication by means other than the application of logical categories, Rahner was nevertheless anxious to defend the process against any accusation of irrationality. To achieve this, he sought to broaden the understanding of 'rational'. Thus, he suggested that 'rational' could be predicated not only of what was derived from a syllogism, but also of the level of certainty that could be attached to convictions such as, 'God exists', and 'the human person is free rather than determined'.[15] As an example of the 'total, primordial, and absolutely rational certainty' which accompanied such convictions, Rahner offered the assurance attached to the belief that, if in their right minds, our mothers will not poison us.[16] From this basis, he argued that the Church too could claim a similar certainty, a similar rationality, for the method employed in the development of dogma:

Why should the work of reflective theology not be considered as the necessary, though always inadequate reflection of the faith consciousness of the Church on a primordial rational certainty by which it knows the connection between an old and new utterance of dogma? Why should such theological labour (supposing it has been carried out with the utmost strictness and honesty, but also in an inner, sympathetic contact with the object and the primordial truth) not be qualified as certain knowledge, if one does not forget that this certainty lives by reference to

[14] Ibid. 61 (ibid. 73). Rahner's emphasis. As suggested by the translator, in a fn. also to be found on p. 61, Rahner's insertion of a hyphen between *mit* and *geteilt* is designed to emphasize the inclusive nature of God's communication, the fact that God actually intends all the meanings which could be derived from God's communication.

[15] 'Considerations', *TI* iv. 20 (*ST* iv. 32).

[16] Ibid. 21 (ibid. 33).

that more primordial certainty and is always sustained by it, that rational reflection does not create that certainty for the first time, since it rather lives by that certainty?[17]

In the above, what has been outlined is Rahner's understanding of why logical principles alone could not be the means by which the Church grew in its understanding of revelation. What will be done in the following section is to substantiate Rahner's further claim that the Spirit was the heart of this process of unveiling revelation.

THE SPIRIT: CONTENT AND AGENT OF THE CHURCH'S FAITH

To appreciate why Rahner should have regarded the Spirit as the source of the Church's ability to unfold revelation, it is necessary to return once again to the fact that his definition of revelation as God's self-communication meant that the truth which the Church came to recognize was not a proposition about God, but was God. Consequently, the Spirit—'Spirit of the Father and of the Son, Spirit of the Crucified and Ascended, Spirit of the Church'[18]—was not simply the Church's guide, but was itself that truth which the Church was able to appropriate. This was not, however, the limit of Rahner's claims for the Spirit. As well as portraying the Spirit as the content of the Church's growth in faith, he argued that the Spirit was also the means by which the Church came to understand revelation:

> Knowledge in faith takes place in the power of the Spirit of God, while at the same time, that Spirit is the concrete reality believed. . . . It follows that the object of faith is not something merely passive, indifferently set over against a subjective attitude to it, but simultaneously the principle by which it itself is grasped as object.[19]

Since revelation was God's self-communication, it will be clear that the Spirit could be the content of that revelation; what is not as obvious, however, is how the Spirit could be understood as the means by which the Church came to recognize this self-communication. In order to elucidate the role of the Spirit, what has to be investigated is Rahner's understanding of the connection between the Spirit and human consciousness.

[17] 'Considerations', *TI* iv. 22 (*ST* iv. 34).
[18] 'Development', *TI* i. 51 (*ST* i. 62). [19] Ibid. (ibid.).

Rahner claimed that the role of the Spirit in the Church was misrepresented if the Spirit was portrayed simply as 'the transcendent steersman of dogmatic development'.[20] The Spirit was far more: it was the light of faith which actually formed the believing consciousness of the Church.[21] While he thus united the Spirit to the consciousness of the Church, Rahner also emphasized that the Spirit was not to be portrayed as simply another object of that consciousness:

The light of faith and the impulse of the Spirit, do not permit of being isolated for inspection by a reflexive process in which attention is turned back upon itself and withdrawn from the object of faith. They are the brightness which illuminates the object of faith, the horizon within which it is contained, the mysterious sympathy with which it is understood, and not properly the object directly regarded, not a sun which we can immediately contemplate.[22]

Since it was itself the 'a priori horizon of consciousness', the grace of the Spirit could neither be simply an individual datum of that consciousness nor be known by reflection.[23] Its connection to God's self-communication meant that this horizon—which can be understood in terms of the 'supernatural existential' discussed in Chapter 1—was rather

that inexpressible orientation of humanity in knowledge and freedom, an orientation which does not declare itself like that of each individual object, a definition which is at the same time silent and therefore all the more emphatically all-embracing and effective in everything, it is nameless and by that very fact present to everything and above everything specifiable; it is the dynamism of the spirit's transcendence into the infinity of the silent mystery we call God, the dynamism which is really meant to arrive and to accept, and not merely to be the eternally asymptotic movement towards the infinity of God; it is meant to reach the infinity of God since God gives God's self to it of God's own accord . . .[24]

Thus, since it was both the light of the Church's faith and also the truth that the Church sought ever more deeply, the Spirit enabled the Church to know by connaturality, as in the case of

[20] 'Considerations', *TI* iv. 14 (*ST* iv. 24).
[21] Ibid. (ibid.). [22] 'Development', *TI* i. 51 (*ST* i. 62).
[23] 'History of the World and Salvation History' ['History'], *TI* v. 103 ('Weltgeschichte und Heilsgeschichte', *ST* v. 122), orig. pub. 1963.
[24] Ibid. 103–4 (ibid. 122–3).

the formation of the Canon of Scripture, what was in accord with its nature and, therefore, genuinely revelatory.[25] Since the Spirit unfolded for the Church the mystery of the triune God, this meant that the development of the Church's faith was not to be measured solely by the proliferation of individual assertions about God, but also by a deepening awareness of 'the blessed darkness' of the mystery of this God.[26] Consequently, the development of dogma tended to compression and simplification rather than to multiplication.

Rahner's stress on the Spirit's activity in the Church should not be interpreted, however, as tantamount to a rejection of the importance of the juridically constituted aspects of the Church which were introduced in the previous chapter. Indeed, Rahner's sacramental approach meant that the Spirit was best perceived as animating, rather than opposing, the structures of the Church. Furthermore, the indispensable role of the Church's magisterium in unveiling revelation implied that it enjoyed a unique relationship to the Spirit as the source of the Church's truth. The next task in developing Rahner's view of the Church's charismatic dimension is, therefore, to demonstrate his understanding of this link between the Spirit and the Church's hierarchy.

THE SPIRIT AND OFFICE IN THE CHURCH

Although 'the institutional' and 'the charismatic' are often portrayed as mutually exclusive, Rahner's linkage of the two indicated neither an ignorance of political realities nor recourse to a pious hope. Rather, as has already been indicated, it flowed directly from his conviction that the ability of office-bearers in the Church to fulfil their responsibility of determining an authentic unfolding of revelation required the particular guidance of the Spirit. Office in the Church was, therefore, as Rahner stressed in *The Dynamic Element in the Church*, both institutional and charismatic.[27]

If such a claim seems extraordinary, even more remarkable is Rahner's assertion that, since this assistance of the Spirit could not be reduced to juridical terms, office-bearers in the Church

[25] *Inspiration*, 70 (*Schriftinspiration*, 75).
[26] 'Considerations', *TI* iv. 26 (*ST* iv. 39).
[27] *Dynamic Element*, 42–3 (*Das Dynamische*, 38–9).

could not be made subject to laws and norms.[28] Consequently, there could be in the Church no constitution with dominion over the office-holders, no revolutions justifiable on the grounds that those in office had exceeded their statutory authority.

The fact that office in the Church was the recipient of a particular charism which was not itself subject to juridical control inevitably raises the spectre that those in authority might use this freedom to fashion the Church according to their private whims. Ensuring that this nightmarish possibility never became a reality, was, claimed Rahner, also an element of the Spirit's mission. Since the assistance of the Spirit was given to enable those in office to lead the Church to that truth which was God, it was, therefore, the Spirit alone who could ensure that the hierarchy of the Church did not use its power to rebel against God.[29] Furthermore, Rahner argued that as the papacy was eminently authoritative in the Church, the activity of the Spirit *vis-à-vis* the Pope was necessarily distinctive:

> the Pope also has the competence to determine what his competence is [Kompetenz der Kompetenz], that is, where he invokes his supreme and ultimately binding authority, one cannot oppose his decision with the claim that he has exceeded his competence and that therefore his decision is not binding. That he has acted within the sphere of his competence cannot be proved by a norm which one can call upon as a judge to evaluate his action, rather, when he calls upon his ultimate authority, his action itself is the only guarantee that he has acted within his competence. This means, however, that in order for such an office held by a human being not to become an absolute tyranny . . . it must necessarily be charismatic. That is, such an office is only thinkable if there is always added to it, in fact and in theory, a power which is itself infallible: the assistance of God's own Spirit, who is permanently promised to it, without being able to be administered or juridically constituted.[30]

Rahner's portrayal of the teaching authority as uniquely gifted by the Spirit to articulate the message of divine revelation inevitably determined his assessment of the relationship between the magisterium and the individual believer. As will be seen in what follows, it specifically shaped his understanding of the response due from believing members of the Church to magisterial teachings.

[28] *Dynamic Element*, 44 (*Das Dynamische*, 40).
[29] Ibid. (ibid.). [30] Ibid. 45 (ibid. 40–1).

THE MAGISTERIUM AND THE BELIEVER

In order to understand correctly Rahner's view of the status of magisterial teaching, it is necessary to recall an aspect of his ecclesiology which was examined in both Chapters 1 and 2: the fact that faith was related not simply to Christ, but also to the Church. That Rahner held unequivocally to this conviction is clear in the following:

> Faith does not only mean accepting what 'I' as an individual believe that I have heard. It also means accepting what the Church has heard, giving my assent to the 'confession' of the Church, the Church which is not only the bearer of the message of Christ which it delivers to individuals (and which then disappears again like a postman), but it is the enduring and abiding medium of faith, that in which faith is posited in order that one voice of praise of the living God may resound as from one mouth and one body to give glory to his compassion.[31]

So insistent was Rahner on the priority of the faith of the Church that he argued, for example, that heresy was not to be understood as a simple disagreement on some point of doctrine—the Church can have its opinion, but I can also have mine—but as apostasy from the Church itself and from that insight into the mystery of God which properly belonged to the Church alone.[32] This apostasy from the Church threatened the individual's connection to the original event of revelation, a connection which was essential if belief was to be authentically Christian.[33]

Since the Church enjoyed the particular guidance of the Spirit, it could be described as the 'mistress' of the faith of its members.[34] The Church was, therefore, the measure of faith, rather than that which was measured.[35] While he acknowledged that the Church's status as the standard and mistress of faith resulted entirely from the activity of the Spirit, rather than from any particular virtues of those within the Church, Rahner also claimed that its position

[31] '"I Believe in the Church"' ['I Believe'], *TI* vii. 109–10 ('"Ich glaube die Kirche"', *ST* vii. 112). Although this article did not appear in its present form until 1966, it was orig. pub. in *Wort und Wahrheit*, 9 (1954), 329–39, as a response to the formal declaration of the doctrine of the Assumption. With the exception of the omission of brief refs. to the Assumption, which appeared at the beginning of the orig. work, the two articles are identical.

[32] Ibid. 110 (ibid.).

[33] 'What is Heresy?' ['Heresy'], *TI* v. 469–70 ('Was ist Häresie?', *ST* v. 528–9), orig. pub. 1961.

[34] 'I Believe', *TI* vii. 112 (*ST* vii. 114).

[35] Ibid. 110 (ibid. 113).

was none the less 'unconditional'.[36] The corollary of the magisterium's exclusive competence to articulate the faith of the Church was the duty of the believer to accept its teaching. In addition, the Spirit's guidance of the teaching office in the Church meant that there could be no appeal to an authority higher than the magisterium. As a result:

> When the teaching authority of the Church requires the assent of faith to some defined doctrine, then there are no precautions or safety measures (whether in the writings of the Church or in any a priori established methods of theology) on which one could call to ensure that the Spirit of the Church is not making any mistake.[37]

Although the above might give the impression that Rahner's approach to the obedience due to the magisterium was totally lacking in nuance, such was not the case. He affirmed, for example, that the infallibility of the papal teaching office applied only under clearly defined conditions in the context of faith and morals. In addition, he also argued that while it belonged to the teaching authority alone to define the faith of the Church, any formal definitions of that faith usually crowned the efforts of 'the unofficial elements of development'—the charisms at work in theologians—rather than those of the magisterium itself.[38] Furthermore, he recognized that, in its pastoral practice, administration, and response to the needs of a particular age, the Church's hierarchy could certainly be characterized by 'error, neglect, partially false developments, the appearance of paralysis, and reactionary tendencies'.[39] Nevertheless, Rahner was also at pains to stress, in a way reminiscent of his view of the sinful Church, that portraying office in the Church as an amalgam of an inspired teaching authority and an otherwise unholy life was contrary to the general holiness of the Church, which itself derived from the Spirit.[40] In addition, he emphasized that even when those involved in proclaiming the Church's belief were sinful and narrow-minded, this did not prove that the Spirit was absent from their teaching.[41]

The preceding examination of the work of the Spirit in the Church has concentrated on Rahner's understanding of the

[36] Ibid. 112 (ibid. 114). The pub. trans. makes the point even more strongly by rendering *unbedingt* as 'absolute and unconditional'.
[37] Ibid. 118 (ibid. 119). [38] 'Considerations', *TI* iv. 15 (*ST* iv. 26).
[39] *Dynamic Element*, 47 (*Das Dynamische*, 42). [40] Ibid. (ibid.).
[41] 'I Believe', *TI* vii. 118 (*ST* vii. 120).

charisms of office. The characteristic feature of his approach to the charismatic in the Church was, however, its universality. Thus, he not only avoided portraying the Pope and the bishops as the sole recipients of the Spirit's guidance, but he also explicitly acknowledged that being without office did not mean being without gifts which could contribute to the life of the Church. The study of the nature and exercise of such gifts can profitably begin by exploring his stress on freedom and individuality in the Church.

FREEDOM, INDIVIDUALITY, AND THE CHURCH

For Rahner, the key to understanding freedom was the awareness that it was a theological notion. Freedom was not, therefore, to be equated either with choosing between limited possibilities—which might entail merely being able to select in which corner of a prison cell to stand—or with a private possession to be jealously guarded.[42] If human freedom had involved no more than a choice between finite objects, it would have been only relatively valuable. As interpreted by Rahner, however, freedom was absolutely valuable. It was so because freedom meant the opening of the person to what transcended human limits. This was possible, of course, only if this freedom was derived from, and directed towards, God.

As has been indicated in discussing both his notion of the supernatural existential and his theology of revelation, Rahner emphasized that our experience of God not only took place in history, but also was mediated by the objects of our experience. The corollary of this approach was that our response to God also had to occur in history. This response took place via our concrete choices, those choices whereby we disposed of our freedom. Fundamental to Rahner's hermeneutics of freedom was the belief that God was 'the supporting ground and ultimate orientation' of every act of human choice. This 'unthematic' presence of God wherever human beings made choices meant that the freedom exercised in those choices could not be understood without reference to God.[43]

[42] 'Freedom in the Church' ['Freedom'], *TI* ii. 91–2 ('Die Freiheit in der Kirche', *ST* ii. 98–9), orig. pub. 1953.

[43] 'The Theology of Freedom', *TI* vi. 180 ('Theologie der Freiheit', *ST* vi. 217), orig. pub. 1964. The original reads: 'weil in jedem Akt der Freiheit Gott als ihr tragender Grund und letztes Woraufhin unthematisch gegeben ist'. See also *Hearers*, 105–6 (*Hörer*, 114–15). The background of Rahner's theology of freedom is discussed by Bonsor, *Rahner, Heidegger, and Truth*, 89–116.

Consequently, all of our choices—not simply those pertaining to the specifically 'religious'—either brought us closer to God or distanced us from God. In other words, the particular expression of freedom constituted by being able to choose between objects, to choose whether to act in this way or that way, also involved a choice for or against God. Our orientation towards God meant, however, that we could never be completely satisfied by any finite objects.[44]

Since God was the source of human transcendence, it followed that God was also the proper object of that freedom. Ultimately, therefore, freedom was the freedom to choose to accept or reject the capacity to transcend the finite; the freedom to choose to accept or reject God. From his anthropological basis, Rahner was able to claim that the obligation to exercise this freedom was not exclusive to the religious person, but was an irreducible element of what it meant to be human:

> Freedom never happens as a merely objective exercise, as a mere choice 'between' individual objects, but is the *self*-exercise of the people who choose objectively; only within this freedom in which people are capable of achieving themselves are people also free with regard to the material of their self-achievement. People can do or omit this or that in view of their own self-realisation that is inescapably imposed on them. This self-realisation is a task they cannot avoid and, in spite of all the differences within the concrete material of this self-achievement, it is always either a self-realisation in the direction of God or a radical self-refusal towards God.[45]

In addition, since human freedom was not simply a transcendental notion, it had a history, a history which was centred on Christ, in whom humanity had been reconciled to God, 'the freedom of our freedom'.[46] Without Christ, human beings would have been unable to transcend their finiteness. Without Christ, freedom

[44] 'Theology of Freedom', *TI* vi. 179–80 (*ST* vi. 216–17).

[45] Ibid. 185 (ibid. 224). Rahner's emphasis. The positive implications for ethical theory of Rahner's anthropology are developed by James F. Bresnahan in 'Rahner's Ethics: Critical Natural Law in Relation to Contemporary Ethical Methodology', *Journal of Religion*, 56 (1976), 36–60. A more general survey of the contribution of Rahner's anthropology to moral theology—a survey which also identifies and answers some criticisms of Rahner's method—is Ronald Modras's 'Implications of Rahner's Anthropology for Fundamental Moral Theology', *Horizons*, 12 (1985), 70–90.

[46] 'Freedom', *TI* ii. 94 (*ST* ii. 100).

of choice would have served only to multiply sin by limiting human beings to the pursuit of self-interest. Through Christ, the absolute saviour, the grace of freedom—and the concomitant possibility of eternal salvation—had become constitutive of human existence.

Since the grace of freedom was an expression of God's action in Christ, its presence in history was symbolized by the Church, the sacrament of Christ. The fact that the Spirit was the 'animating entelechy' of the Church meant that the Church was the site of the freedom bestowed by the Spirit.[47] Belonging to the Church was, therefore, a call to freedom. For this reason, Rahner recognized the need for the Church to guard against the danger of becoming 'a clerical, religiously camouflaged form of a totalitarian system'.[48] The Church had to become in appearance what, through the indwelling of the Spirit, it was essentially: the antithesis of totalitarianism.

In Rahner's analysis, the corollary of freedom was individuality. Without denying their shared materiality, Rahner stressed that each person, as a result of God's unique creation, also reflected the perfect individuality to be found within the Trinity, an individuality which simultaneously embraced community.[49] Thus, far from simply being a member of a group unable to be differentiated from fellow members, every human being could be understood as both an irreplaceable spiritual personality and also as being capable of forming a community with other unique beings.[50] Consequently, the Church itself could be defined as: 'the supernatural community of unique non-interchangeable individuals in interior grace, in the truth and love of Christ, in the self-communication of the three-personed inner life of God'.[51]

There was, therefore, a need for this individuality to be maintained even within the Church. Indeed, immediately after the Second World War, Rahner cautioned all members of the Church against succumbing to a contemporary trend which expressed itself

[47] 'Freedom', *TI* ii. 97 (*ST* ii. 103). [48] Ibid. 99 (ibid. 106).

[49] 'Individual', in *Nature*, 55–6 (*Gefahren*, 15).

[50] Ibid. 57 (ibid. 16). The capacity of Rahner's understanding of freedom to support, simultaneously, both autonomy and community is assessed positively by Robert L. Hurd, 'The Concept of Freedom in Rahner', *Listening*, 17 (1982), 138–52.

[51] 'Individual Member', in *Mission*, i. 135 (*Sendung*, 102).

in 'a willingness to go with the group and an anti-individualistic diffidence'.[52]

At the beginning of this chapter it was claimed that Rahner's understanding of the role of the Spirit in the Church allowed him to combine elements which, at first glance, would seem to be contradictory. Thus, he could affirm the need not only for continuity, but also for development in the articulation of the Church's faith, and he could acknowledge that both the hierarchy of the Church and believers generally shared in the gifts of the Spirit. In the following section, this ability to hold opposites in tension will be illustrated in the context of Rahner's commitment to both individual freedom and the right of the Church to make decisions binding on all its members.

INDIVIDUAL FREEDOM AND UNIVERSAL MORALITY

In a series of articles written during the 1940s and 1950s, Rahner developed a characteristically measured approach to the issue of freedom of conscience. To begin, he affirmed that, since each of the baptized was a 'materially, biologically individuated member of a species', there could be within the Church, viewed as an organized society, some laws that applied universally.[53] Consequently, where it was possible to introduce universal norms—such as the prohibition against direct abortion or the indissolubility of sacramental marriage—there could be no appeal against them on the grounds of individual conscience or circumstances.[54] He argued that promoting 'situation morality' in contexts where universal norms were applicable was an ethical and metaphysical nominalism.[55] Indeed, he castigated situationism generally for isolating consideration of the morality of actions from any regard for the person as an ethical and religious subject.[56]

Rahner claimed that when the teaching office authoritatively stated that something was irreconcilable with the will of God, this was to be preferred to the human capacity for self-deception,

[52] 'Individual', in *Nature*, 53 (*Gefahren*, 13). [53] Ibid. 65 (ibid. 23–4).

[54] 'The Appeal to Conscience' ['Conscience'], in *Nature*, 99–100 ('Der Appell an das Gewissen', in *Gefahren*, 52), orig. pub. 1949.

[55] Ibid. (ibid.).

[56] 'Der Anspruch Gottes und der Einzelne' ['Anspruch'], *ST* vi. 521–2. This article, which was orig. pub. in 1960, was never included in any vol. of *TI*.

which could subvert even freedom of conscience.[57] Significantly, Rahner emphasized that the binding nature of the Church's teaching authority in such cases applied irrespective of whether such teaching came from the 'ordinary' or 'extraordinary' magisterium.[58] In addition, he argued that the obligation to accept the Church's teaching that particular actions were incompatible with Christian morality was not dependent on the believer's understanding why a particular proscription had been introduced. Where this comprehension was lacking, the faithful Christian had to be satisfied with the fact that the Church's legitimate authority, whose insight was greater than that of the individual, had made the decision.[59]

The Church's action in ruling that certain things were immoral under any circumstances was not viewed by Rahner as a denial of freedom. He argued that such bans, which were directed only against individual acts, did not impair freedom, because their proscriptions were limited to those things which, since they themselves were injurious to the dignity of the person, could not actually be listed among authentic objects of freedom.[60] In Rahner's analysis, freedom did not imply an absolute right to error or to objectively false moral behaviour.[61] Furthermore, he claimed that since the restrictions imposed by the Church were binding only on those who, in freely choosing to belong to the Church, recognized that the Spirit worked through its teaching authority, this too meant that the Church's proscriptions were not at odds with the rights of its members.[62] In answer to those who regarded the Church's moral norms as old-fashioned, Rahner argued that the Church's reliance on the Spirit's guidance meant that it was far from self-evident that the Church could be convicted of being out of touch with either 'reality' or 'the times'. For these reasons, he saw no contradiction between obedience to the law and freedom.[63]

As he had combined advocacy of the absolute right of the teaching authority to define doctrine with the recognition that the impetus for such development of doctrine usually came from those outside the hierarchy, so Rahner's defence of the right of the

[57] 'Conscience', in *Nature*, 101–2 (*Gefahren*, 53–4).
[58] Ibid. 98–9 (ibid. 51). [59] 'Freedom', *TI* ii. 102 (*ST* ii. 109).
[60] Ibid. 101–2 (ibid. 108–9). [61] Ibid. (ibid.).
[62] Ibid. 102 (ibid. 109).
[63] 'Conscience', in *Nature*, 103–5 (*Gefahren*, 55–6).

Church's teaching authority to propound universal norms did not prevent him endorsing the need for individual moral decisions. Thus, he argued that, since the Church's moral norms pertained exclusively to what all human beings shared, rather than to what was applicable only to an individual or a specific group, there always remained an irreducible zone of freedom in moral decision-making.[64] In addition, the fact that the Church authoritatively stated only what could not be sanctioned as authentic Christian morality—as distinct from making a parallel determination of what was legitimate—meant that the task of deciding, in each situation, what was in accord with faithfulness to the Spirit remained with the individual.[65] Faithfulness to the Spirit—holiness—demanded, therefore, more than merely avoiding breaches of the Church's universal norms:

> [the Spirit] is not to be found apart from the letter of the New Covenant, but not everyone who recognises this letter as sacred and says 'Lord, Lord' is ipso facto a Christian filled with the Spirit, spiritual in the sense that God, our own responsibility and our own age demand that we should be spiritual. Only he or she who is a member of the Church *and* independent, humble, *and* daring, obedient *and* conscious of his or her own personal responsibility, a pray-er *and* a doer, adhering to the Church in its past *and* in its future—only such a one as this makes room for the Spirit of God at Pentecost . . . for this Spirit to do its work in him or her.[66]

Thus, even after the Church's moral teaching was accepted, there remained a vast range of moral questions which could be answered only in the concrete situation of each individual. Indeed, Rahner emphasized that even if the individual wanted to abrogate the responsibility of such choices, this freedom could not be surrendered to the Church.[67] The Church could do no more than preach a formal kind of individual morality which exhorted people to become what they could be.[68] Since the mystery of spiritual individuality was unfathomable, it was 'above' the Church's competence to specify what path to holiness each person ought follow.[69] In fact, Rahner argued that the individual, by

[64] 'Freedom', *TI* ii. 105 (*ST* i. 112). [65] Ibid. 103 (ibid. 110).
[66] 'Spirit', *TI* vii. 190–1 (*ST* vii. 187). Rahner's emphasis.
[67] 'Individual', in *Nature*, 74–5 (*Gefahren*, 31–2).
[68] Ibid. 72 (ibid. 29–30). [69] Ibid. 70–1 (ibid. 28).

courageous decision-making, could often provide a charismatic example for the Church itself, since the latter was often so burdened by general principles that it failed to respond to concrete issues which called for specific imperatives.[70]

Faithful to the principles of Ignatius of Loyola, Rahner claimed that there was no substitute for the discernment of spirits, by which graced individual sought to identify what was in accord with God's will for them alone.[71] Rahner emphasized that, since each person was an *individuum ineffabile*, Christian conscience had not only to apply universal norms, but also to grasp what was required of the individuals in their unique circumstances: 'The concrete moral act is more than just the realisation of a universal idea happening here and now in the form of a case. The act is a reality which has a positive and substantial property which is basically and absolutely unique.'[72]

Not only the unique history of each individual, but also changing historical circumstances suggested the need for individual decisions. This was particularly so in the twentieth century, when human planning and decisions were constantly bringing about situations and possibilities which could not be foreseen by universal norms.[73] Significantly, writing in 1965, Rahner included the possibility of birth control as among the contemporary developments which fell outside the scope of existing norms.[74]

Rahner's concern with the charisms of the non-ordained was not limited, however, to championing the inviolable right of individuals to pursue their unique paths to holiness. Since each member of the Church, precisely as a member of the Church, was gifted by the Spirit, it followed that all the members of the Church, rather than just the ordained, had a role to play in fulfilling the mission of the Church to make God present in the world. How Rahner understood this mission of the laity will be examined in what follows.

[70] 'Anspruch', *ST* vi. 532–3.
[71] 'Individual', in *Nature*, 77 (*Gefahren*, 33).
[72] 'On the Question of a Formal Existential Ethics', *TI* ii. 225 ('Über die Frage einer formalen Existentialethik', *ST* ii. 236), orig. pub. 1955.
[73] 'Situation Ethics in an Ecumenical Perspective', in *The Christian of the Future* [*Christian of the Future*], trans. W. J. O'Hara (London, 1967), 44–5. Together with three other pieces from *ST* vi, this article, which was orig. pub. in 1965, was never included in *TI*. ('Zur "Situationsethik" aus ökumenischer Sicht', *ST* vi. 541).
[74] Ibid. 45 (ibid. 542).

THE LAITY AND THE MISSION OF THE CHURCH

As part of his commitment to promoting the charisms of the non-ordained, Rahner set himself the task of trying to overcome the pejorative tone associated with 'laity'. Accordingly, in his 1955 article 'Notes on the Lay Apostolate', he argued that the proper theological meaning of 'lay' needed to be distinguished from the popular tendency to regard the word as equivalent to 'profane' or 'ignorant'.[75] He was, however, not unaware how deeply this tendency was ingrained even within the Church itself, where 'lay' was generally perceived as synonymous with the non-expert, with being excluded from the dignity and function of office.[76] Indeed, he recognized that such negative perceptions of the non-ordained had a long history. Since the patristic era, he argued, the Church had become a Church of 'mere institutionalism', a clerical Church in which the laity were not accepted as co-workers.[77]

While conceding that efforts to alter this situation had been in train since the second decade of the twentieth century, Rahner did not naïvely imagine that attitudes which had prevailed for a millennium and a half could be quickly reformed.[78] Nevertheless, he was convinced that the key to amending the position of the non-ordained was the development of a new understanding of 'laity'. In the light of the history of the Church in the last generation, it is worth noting his prayer that such an awareness emerge organically, rather than becoming unavoidable as history reduced the Church to such a 'little flock' that even the clergy–laity division would no longer be tenable.[79]

As a more creative alternative, Rahner promoted the idea that the laity were best portrayed as those who were consecrated and called in to the 'ecclesia' of God: 'the notion of the "lay person", therefore, does not mark the boundary between the sphere of the profane and the sphere of the sacred and sacral, rather it refers to someone who has a *definite* position *within* the one consecrated realm of the Church'.[80]

Rahner's positive view of the laity had its foundation in his understanding of the ramifications of baptism. What was important

[75] 'Apostolate', *TI* ii. 319 (*ST* ii. 339).
[76] 'Layman', *TI* viii. 53 (*ST* vii. 332). [77] Ibid. 52 (ibid. 331).
[78] Ibid. (ibid. 330–1). [79] Ibid. 53 (ibid. 332).
[80] 'Apostolate', *TI* ii. 319 (*ST* ii. 339). Rahner's emphasis.

about baptism was not that it related to individual salvation by opening the way to the reception of the other sacraments—that this was not primary for Rahner can be divined from his conviction, discussed in the Chapter 1, that salvation was possible for all via faith and love alone—but that, through it, each person shared the vocation of the Church.[81] While accepting that, unlike priestly ordination, the 'character' conferred by baptism did not give the individual a new place in the world, Rahner maintained that it did nevertheless give each Christian a new task in the same place: that of making the Church present.[82] His stress on the mission and gifts given through baptism can be found as early as 1936, when, as was mentioned in the previous chapter, he wrote that baptism gave each person a mission of pastoral care.[83] The baptized Christian was to be the bearer of the Word, the witness to truth, and the representative of the grace of Christ in the world.[84] Although only those who held office had the particular mission of guiding the Church, each member of the Church had a role to play if the Church was truly to be the sacrament of Christ in history.

Through the Spirit received at baptism, the laity had a mission to perform precisely as laity. They did not, therefore, have to be transformed into clerics before they could act. The contribution of the lay person to the mission of the Church could not, therefore, be restricted to supporting the Church financially, voting as directed, or simply being involved in pious exercises. Rather, being a baptized Christian

> implies an awareness, so deep and so radical that it revolutionises everything, of the fact that baptised persons are constantly confronted with the task of a Christian in that environment in which they find themselves and in which their lives are passed, that is to say in the natural context of their calling, in their families, the circles in which they live, their nation and state, their human and cultural milieu. And this task consists in establishing the dominion of God in truth, in selflessness and in love, and thereby making what is truly essential to the Church's nature present in the setting in which they are placed, from the position which only they can occupy, in which they cannot be replaced by any other, not even by the clergy, and where, nevertheless, the Church must be.[85]

[81] 'Layman', *TI* viii. 55–6 (*ST* vii. 334–5).
[82] 'Apostolate', *TI* ii. 324–5 (*ST* ii. 344–5).
[83] 'Care of Souls', *TI* iii. 272 (*ST* iii. 323).
[84] 'Layman', *TI* viii. 58 (*ST* vii. 337).
[85] Ibid. 62 (ibid. 340).

The fact that the charisms of the laity, charisms essential to the mission of the Church, originated directly from the Spirit, not only implied that the laity were more than mere appendages to the work of the Church; it also had ramifications for the relationship between the hierarchy and those without office. It meant that

> in the Church to which charismatic elements belong, subordinates are not simply those who have to carry out orders from above. They have other commands as well to carry out: those of the Lord, who also guides God's Church directly and does not always in the first place convey God's commands and promptings to ordinary Christians through ecclesiastical authorities, but has entirely reserved for God the right to do this directly in a great variety of ways that have little to do with keeping to the standard procedure and the 'usual channels'.[86]

Thus, the relationship to the Spirit, which was the patrimony of all baptized Christians, not only commissioned all members of the Church to make Christ present in the situation unique to them, but also had implications for life within the Church itself. Rahner's understanding of these implications will be explored in the next section.

PUBLIC OPINION AND DEMOCRACY IN THE CHURCH

In an address—significantly, to journalists—in February 1950, Pius XII remarked that 'public opinion' was as necessary in the Church, for both its shepherds and flock, as in any society.[87] Throughout the 1950s and early 1960s, Rahner often referred to this comment in urging that the laity not be denied the opportunity to make their contribution to the life of the Church.[88]

Clearly, however, Rahner did not interpret the need for public opinion as a suggestion that the Church ought to be run by opinion polls. Indeed, he argued that the charisms of the Spirit given to the laity meant that public opinion in the Church was to be understood on 'a higher level' than in secular usage.[89] These

[86] *Dynamic Element*, 70 (*Das Dynamische*, 62).

[87] *L'Osservatore Romano*, 18 Feb. 1950.

[88] Rahner's ref. to Pius XII's comment can be found in: *Free Speech in the Church* [*Free Speech*], trans. G. R. Lamb (New York, 1959), 14–15 (*Das freie Wort in der Kirche* [*Wort*] (Einsiedeln, 1953), 8–9); 'On Conversions to the Church', *TI* iii. 383 (*ST* iii. 452); 'Belief Today', *TI* v. 16 (*ST* v. 25–6).

[89] *Free Speech*, 19 (*Wort*, 12).

charisms meant that the 'popular element' in the Church was not simply 'the people of the earth', but 'the people of God'.[90] This fact alone meant that the impact of the voice of the people within the Church ought to be greater than in secular society. This voice was not, therefore, to be restricted to the periphery of the Church's life.

Indeed, Rahner stressed that those without office, those who constituted 'the listening Church', could actually help the Church grow in the truth, since they too were guided by the Spirit, who was both the source and content of that truth.[91] Unless the faith of the 'believing Church', which actually included both the hierarchy and the laity, was 'more'—understood in terms of its intensity, rather than its quantity—than that of the teaching Church alone, there could be no possibility of the development of doctrine.[92] This faith, which could never be exhausted by dogmatic statements, was always growing, a growth which came about through an ever-deeper acceptance of God's self-communication.

While the hierarchy had a particular responsibility in regard to the formulation of the Church's faith, there were many other areas in the life and mission of the Church where they could claim no special gifts. Thus, neither bishops nor priests were infallible, nor had a monopoly of charisms, in such matters as pastoral care, piety, and liturgy. The fact that such spheres were not only fundamental to the Church's life, but also challenged the Church's practice to develop to meet changing historical conditions, meant, argued Rahner, that office-holders ought to accept guidance in them from those who were immersed in that ever-evolving social context: without the aid of such public opinion, the decisions of the hierarchy could fail to respond to genuine needs.[93]

In the light of Rahner's staunch advocacy of the need for public opinion not only to exist in the Church but also to be heard, it can be asked whether his approach amounted to an effort to democratize the Church. Although the issue of whether, and to what extent, the Church ought to be a democracy has become something of a shibboleth in the skirmishes between contemporary liberals and conservatives, Rahner approached the question in theological rather than ideological terms. That his method involved

[90] *Dynamic Element*, 72–3 (*Das Dynamische*, 64).
[91] *Free Speech*, 18 (*Wort* 12).
[92] 'Ecclesiological Piety', *TI* v. 347–8 (*ST* v. 391).
[93] *Free Speech*, 19–26 (*Wort*, 13–18).

neither renouncing his belief that the Church's hierarchical structure existed *iure divino* nor advocating a popular plebiscite on every issue can clearly be seen in the following, which was written in the year after Pius XII's defence of public opinion in the Church:

[The Church's] authority is not from the people, but from the grace of Christ. It derives ultimately not from a vote from below but from investiture from above; the laws governing its action are given in its permanent and unchangeable constitution, which was established for it by its Lord; the Church is in its real nature, despite all its historicity, development and involvement with external, secular powers, not the product of changing powers of secular history, but the unique and permanent foundation of the Lord till the end of time.[94]

Rahner did, however, seek to combine this conviction with the view that, if democracy was the opposite of a totalitarian system within which all power is assembled in one pair of hands, some features of a democracy could indeed be found within the Church.[95] He interpreted the spirit of democracy as meaning not that everyone had a vote—which, he suggested, could itself introduce the tyranny of the majority—but that there was a necessary plurality of powers, thereby ensuring protection against the excesses of any one of those powers.[96] Within the Church, the source of this plurality was the Spirit, who guided both the laity and the hierarchy. As will be seen in the next section, Rahner himself had a number of proposals, all of them derived from the exercise of that freedom which was the fundamental gift of the Spirit, for what would allow that plurality to be seen unequivocally.

THE PLURALITY OF CHARISMS AND THE CHARISM OF PLURALITY

There were, argued Rahner, three things which were necessary if the Church was to be other than a totalitarian structure. The first was that there be, within the limits determined by what was defined dogma, different schools of theology and freedom of research and opinions.[97] While acknowledging the right of those in authority to adjudicate when such opinions had in fact gone beyond what was consistent with the Church's faith, he also urged

[94] Ibid. 13 (ibid. 8).
[95] *Dynamic Element*, 71 (*Das Dynamische*, 63).
[96] Ibid. 71–2 (ibid.).
[97] 'Freedom', *TI* ii. 107 (*ST* ii. 114).

that these adjudications ought to err on the side of mercy rather than rigorism.[98] In what was perhaps a *cri de cœur* in the light of his own pre-conciliar experiences, he argued that since the theologian acting in good faith actually provided the Church's teaching office with the opportunity to clarify where it stood on non-defined topics, such theologians ought to be treated honourably even when their opinions were ultimately rejected.[99]

Rahner balanced his loyalty to the teaching office with the belief that the charismatic dimension of the Church could also be revealed in the existence of 'Her Majesty's loyal opposition'.[100] He asserted that the Church's history clearly showed that it was often the saints who had expressed divinely-willed opposition to the merely human in the Church. Such faithfulness to the Spirit guaranteed, however, neither earthly vindication nor popularity for those who challenged Church authorities. Consequently, Rahner endorsed Newman's assertion that living 'under a cloud' as far as one's status in the Church was concerned, could often indicate obedience to the Spirit.[101]

While the fear of giving the appearance of disunity often convinced those in authority of the need to restrict freedom of discussion, Rahner rejected such a motive. He emphasized not only that such discussions could be the work of the Spirit drawing the Church to a deeper understanding of truth, but also that repression, whatever its motives, would not enable the Church to hide from the world what the world already knew: that the Church did not have ready-made answers to every question.[102]

Secondly, long before the phrase became a standard part of post-Vatican II jargon, Rahner emphasized the need to strengthen 'the principle of subsidiarity'.[103] This meant that, without prejudice to the *ius divinum* nature of authority in the Church, not everything had to originate from the centre. If all the members of the Church were gifted by the Spirit, office-bearers were not to act as if administering a centralized bureaucracy where individuals

[98] *Free Speech*, 27 (*Wort*, 18–9). [99] Ibid. 29–30 (ibid. 20).
[100] Ibid. 37 (ibid. 26). [101] Ibid. (ibid.).
[102] 'Individual Member', in *Mission*, i. 168 (*Sendung*, 124–5).
[103] Rahner's first ref. to this principle was in his 1948 article 'Peaceful Reflections on the Parochial Principle' ['Parochial Principle'], *TI* ii. 303 ('Friedliche Erwägungen über das Pfarrprinzip', *ST* ii. 321). In that article, he acknowledged that the expression had in fact been used by Pius XII as early as 1946. See also the 1953 article, 'Freedom', *TI* ii. 105 (*ST* ii. 112).

were allowed no initiative.[104] The alternative promoted by Rahner was a Church in which a 'freedom of association' made possible initiatives from below.[105] A bureaucratic 'State socialism' which sought to suppress such movements could lead only to the death of real life in the Church.[106] A corollary of trusting in the Spirit's guidance of the Church was, therefore, accepting that neither papal primacy nor the centralized bureaucracy which derived from it was the Church's only guarantee of truth.[107]

The right of individuals to respond to their charisms did not, however, apply only to the laity. As was discussed in the previous chapter, Rahner also stressed the need for local bishops to exercise a similar creativity. While the approval of the Pope could indeed be an indication that the bishop was led by the Spirit, each bishop was more than simply a channel for what came from Rome.[108] Each bishop was, therefore, obliged to openness to impulses from the Spirit and not simply from the Curia.

Thirdly, the necessary freedom of theology and the principle of subsidiarity both required that those in authority exercise patience and charity towards the existence of non-hierarchical charisms. The responsibility of office-bearers to preserve the faith of the Church was not, therefore, tantamount to a command to crush, or devise unnecessary difficulties for, what emerged 'from below'.[109] Rahner recognized that new charisms were always shocking and could be mistaken for mere enthusiasm or lack of respect for tradition, but he also maintained that the Church needed the courage to accept that the Spirit could be present in what was new.[110] This applied even though bishops could find that charismatic individuals made life difficult for them.[111] For Rahner, the source of unity between the potentially divisive charisms of the hierarchy and the laity was itself the origin of those same charisms: the Holy Spirit.

As understood by Rahner, unity was charismatic rather than institutional.[112] Similarly, only the Spirit, rather than legislation, could so order the relationship of the primacy and episcopate that

[104] 'Freedom', *TI* ii. 105 (*ST* ii. 112). [105] Ibid. 107 (ibid. 114).
[106] Ibid. (ibid.). [107] 'Stifle', *TI* vii. 76 (*ST* vii. 81).
[108] *Episcopate*, 32 (*Episkopat*, 32).
[109] *Dynamic Element*, 80 (*Das Dynamische*, 71).
[110] Ibid. 82–3 (ibid. 73). [111] 'Freedom', *TI* ii. 107 (*ST* ii. 114).
[112] *Dynamic Element*, 52–3 (*Das Dynamische*, 47).

neither excessive centralism nor the decomposition of the Church's unity through the action of a local bishop could become the norm.[113] In addition, legislation could not achieve a synthesis of the virtues of criticism and obedience, both of which were nevertheless essential to the well-being of the Church. Rahner believed that such charisms could in fact coexist only when all those involved sought to fulfil God's will rather than their own.[114] Without such openness to the Spirit, Church authorities could tend to regard every criticism as disloyalty, while those whose views were rejected by Church authorities could interpret such action as a denial of legitimate freedom. In Rahner's analysis, it was spirituality, rather than precepts, which provided the key to unity. Rahner's conviction on this matter can be illustrated by referring to his suggestion, to be found in an article written in 1948, regarding how disputes between advocates of different pastoral strategies could be resolved:

> Norms, no matter how clearly and subtly reasoned out in a legal sense, can never replace a right moral attitude on the part of those who must put them into practice. Frictions arising from the lack of such an attitude can never be avoided by guiding principles and fixed spheres of action, no matter how exactly thought out they may be. Whenever such frictions arise, we must first of all call for an examination of consciences and not for an examination by the jurists.[115]

There was, nevertheless, one area where Rahner in his early ecclesiological works did not disavow a role for Church law in this process of co-ordinating charisms: protecting the legitimate freedom of the laity against encroachment from office-bearers. While accepting that any such laws could only be *ius humanum*, rather than *ius divinum*, he nevertheless believed that if the rights of lay people were made explicit, it would be possible for them to come to a greater consciousness of their duties in and for the Church.[116] He argued that if Rome was not to legislate for the rights of the laity in the whole world, then individual dioceses and countries ought to do so.[117] By guaranteeing freedom, such laws could promote a greater involvement in the life of the Church: 'Real

[113] *Episcopate*, 35 (*Episkopat*, 35). [114] *Free Speech*, 37 (*Wort*, 26).
[115] 'Parochial Principle', *TI* ii. 315 (*ST* ii. 334).
[116] 'Freedom', *TI* ii. 105–6 (*ST* ii. 113).
[117] 'Apostolate', *TI* ii. 329–30 (*ST* ii. 350).

responsibility and duty will only be accepted and borne where the *law* grants a certain realm of freedom (even though merely *iure humano*) for autonomous fulfilment of such duties and responsibilities.'[118]

While it must be acknowledged that Rahner's support for laws to protect the rights of the laity was a departure from his general approach to the Spirit's role of co-ordinating charisms, it did not mean that he had abandoned the Spirit for a litigious mentality. Far from doing so, he stressed that, since charisms were given for the well-being of the whole Church, there could be no justification for forming groups dedicated to supporting a charism which the Church as a whole did not accept.[119] Indeed, he argued that patiently and humbly accepting the suffering involved in seeing a charism rejected could itself be a sign that the charism was genuine.[120]

REVIEW

A distinguishing feature of Rahner's ecclesiology was its ability to hold in tension two potentially divisive poles. This aspect of his work is pre-eminently illustrated in each of the three main topics discussed in this chapter. Thus, Rahner was concerned to show not just that the Spirit was both the means and content of the Church's truth, but that it was also the Spirit who initiated and guided the processes of development by which the Church grew in that truth. Secondly, his identification of the Spirit as the source of human freedom did not issue in an unequivocal declaration of the supremacy of the individual, but in the recognition that human freedom had to be modelled on the Trinity, where individuality coexisted with community. As a result, Rahner could champion both universal moral norms and an irreducible sphere of individual morality. Thirdly, in addition to defending the indispensable leadership role of the hierarchy, Rahner also stressed that the laity were not only gifted by the Spirit, but gifted for the same purpose as the Pope and the bishops: to enable them to contribute towards making the Church a living sign of Christ in the world.

[118] Ibid. 329 (ibid.). Rahner's emphasis. A similar point is made in *Free Speech*, 48 (*Wort*, 35).

[119] *Dynamic Element*, 78 (*Das Dynamische*, 69).

[120] Ibid. 79 (ibid.).

Rahner's ability to link these diverse strands is regarded by some critics as having been purchased at the cost of doing justice to one or other of the elements involved. Thus, for example, Medard Kehl claims that Rahner's desire to avoid an overemphasis on the institutional aspects of the Church resulted in an ecclesiology which, because it situated office in the Church 'far too much' in the context of the whole Church, did not sufficiently emphasize the uniqueness of that office.[121] On the evidence to be found in this chapter, such a criticism seems, however, hard to justify. While it is certainly true that Rahner stressed the dependence of the magisterium on the faith of the whole Church, it is also true that he was unequivocal in his promotion of the prerogatives of the Church's hierarchy. Indeed, his defence of the magisterium's exclusive right to define the faith of the Church was, as has been noted, a fundamental aspect of his ecclesiology.

As is often the fate of those who try to negotiate a path between two extremes, Rahner's ecclesiology has been attacked by both sides of the institutional–charismatic divide. Thus, Kehl's charge that Rahner failed to highlight the institutional aspects of the Church finds its counterpart in Leo Dullaart's allegation that Rahner did not provide a concrete guide to the relationship between the charisms of the hierarchy and the charisms of those not holding office. As a result, suggests Dullaart, the only possible outcome of any conflict between office and charism would not be the victory of those with the greater conviction, but of those able to combine conviction with power.[122]

Dullart's criticism highlights an underdeveloped area in Rahner's work. Indeed, Rahner often gave the impression that once the Spirit is summoned, human conflicts dissolve in a magical way. A just evaluation of Rahner demands, however, the acknowledgement that Rahner's vagueness on this point reflects not an oversight, but his belief that the work of the Spirit could not be pre-programmed or subjected to invariable norms. None the less, as has been noted, Rahner's support for laws to protect the rights of the laity was at least an indication that he himself recognized that non-institutional charisms had not always fared well in the Church's history. What Rahner did not discuss, however, was

[121] Kehl, *Kirche als Institution*, 222.
[122] Dullaart, *Kirche und Ekklesiologie*, 177.

whether such laws could themselves be interpreted as the work of the Spirit. Even though he may have been vague on the concrete relationship between the charisms of those with and without office in the Church, what emerges clearly from his work is the conviction that a fruitful relationship was possible only if both parties were committed to discerning the movement of the Spirit, and open to be converted by the Spirit. This notion will be important in Chapter 6 in the context of the relationship between theology and the magisterium.

With only minor exceptions, Rahner's treatment of the Church's capacity to change has thus far been presented without reference to how the extent of such change might be affected by particular historical conditions. Accordingly, what has been highlighted is his conviction that change and development belonged to the fundamental character of the Church. Rahner was, however, also aware that the Church existed in history. In addition, he recognized both the speed with which that history was evolving in the twentieth century and the impact it was having on the Church. Indeed, it was his assessment of the challenge which the twentieth century presented to the Church which shaped his ecclesiology in the years immediately prior to Vatican II and, even more dramatically, in the wake of the Council. The next chapter, therefore, will focus on his understanding of the relationship between history and the Church in the twentieth century.

PART TWO

4

'The Shock of the New': The Church and the Twentieth Century

FEW characters in English literature are more haunting than Charles Dickens's Miss Havisham. Jilted at the altar, she refused to accept that time had moved beyond that fateful day. Time, however, was not to be denied. Her downward spiral to insanity, the putrescent mound which the planned wedding-breakfast had become, and her threadbare, but once beautiful, bridal gown all witnessed to the inexorable movement of history. Her experience is a poignant reminder that we live in a world which, despite our best efforts, refuses to stand still. As a result, not only is it impossible to recapture the past, but the attempt to do so produces only eccentricity and incongruity. This principle exempts not even the Church.

Since its constitution as the sacrament of Christ means that the Church exists in history rather than in light supernal, a refusal either to acknowledge the dynamism of that history or to participate in it must imperil the Church's capacity to fulfil its mission. In Rahner's theology, such a refusal was not to be found. Indeed, as the present chapter aims to show, it was Rahner's analysis of the twentieth century which radicalized his approach to change in the Church.

Thus far, this study has concentrated on the theoretical basis of Rahner's understanding of change in the Church. Accordingly, the following have been highlighted: the Church's sinfulness, which provided a motive for reform designed to make the true nature of the Church more evident; the incapacity of Christian doctrine fully to express the mystery of God, which established the need for development of doctrine; the gifts of the Spirit bestowed on all the baptized, which meant that the Church's structures would need reform whenever they ignored or abused non-hierarchical charisms. In addition, Rahner's commitment to both apostolicity and continuity in the Church's doctrine and structures has also been detailed. It was, however, his reflections on what was required if the Church was to respond to the challenge of the twentieth century, the era of 'future shock', which led him to regard change in the Church not merely as a theoretical possibility, but as an urgent necessity.

Rahner's awareness of the Church's links to history did not induce him to write a major theological reflection on the nature of history—although to a certain extent he also provided even that—but rather this awareness was the catalyst of his proposals for major changes in the internal workings of the Church and in its relationship with the world. The inquiry into his interpretation of the twentieth century's impact on the Church can, however, best begin by surveying a further aspect of the theological superstructure for Rahner's understanding of change: the connection between profane and salvation history.

HISTORY AND SALVATION HISTORY

'History of the World and Salvation History', Rahner's reflection on the salvific dimension of history, was first published in 1962—surprisingly late, since his studies of the impact of contemporary society had begun to appear in the 1940s. The basic thesis of that article was that existence in history, when considered in the light of humanity's destiny to be with God, was far more than a negative or even neutral factor. Indeed, Rahner stressed that history had a salvific significance, a significance which derived from its being the sphere where human freedom—which Rahner interpreted as the freedom to choose whether to accept or reject the life-giving self-communication of God—was exercised.[1]

As was indicated in the previous chapter, Rahner believed that the choice for or against God was always a choice made in an encounter with the world, an encounter which was not restricted only to the context of the explicitly religious.[2] While God alone, since salvation was ultimately God's gift, could determine whether our choices were saving or damning—that is, whether they were for or against God—the fact that such choices had necessarily been made in history resulted in a symbiosis between profane

[1] 'History', *TI* v. 98 (*ST* v. 116). [2] Ibid. 98–9 (ibid. 116–17).

history and salvation history.[3] Consequently, Rahner could assert that even the transient and seemingly unimportant events which constituted human history were in fact pregnant with eternity and eternal life or with eternal ruin.[4]

While affirming that the two histories were intertwined, Rahner nevertheless maintained that a distinction between profane and salvation history was possible. This distinction had its source in the Incarnation. Rahner's argument was that, in Christ, God had definitively illuminated a particular part of the otherwise ambiguous history of humanity in order to show where salvation was to be found.[5] Furthermore, the Church, as a result of its foundation in Christ, was to be understood as the site where God's offer of salvation became tangible in the time after the Resurrection:

> Only in Jesus Christ did the divine and human reach an absolute and indissoluble unity; only in the self-revelation of Jesus is this unity also historically possible; only now is this saving history clearly and permanently distinguished from all profane history, and everything, such as the Church, the sacraments and the Scriptures, which follows from this Christ-event and which participates in its own way in this unsurpassable finality of the Christ-event, participates also in its distinction from profane history. Here in Christ and in the Church, saving history reaches its clearest and absolutely permanent distinction from profane history . . .[6]

Rahner asserted that the revelation in Christ established unequivocally that God, not the world, was the source and content of salvation.[7] As a result, humanity could know with certainty that the Kingdom of God would not simply evolve from history. History, encompassed as it was by the inescapable reality of death, would not experience the end of the opposition between man and woman, rich and poor, or war and peace. History would never become, therefore, the place of 'eternal peace and shadowless light'.[8] However, since it remained true not only that our salvation was connected with our choice for or against God, but also that these choices were exercised in history, it followed that Christians were not entitled to withdraw from history. In fact, far from advocating a Christian ghetto, Rahner stressed that Christians were obliged to live in the world in a way that witnessed to their creed

[3] Ibid. 102–3 (ibid. 120–1). [4] Ibid. 99 (ibid. 117).
[5] Ibid. 106–8 (ibid. 125–7). [6] Ibid. 109 (ibid. 129).
[7] Ibid. 111 (ibid. 131). [8] Ibid. 112 (ibid. 132).

that nothing which occurred in history could separate humanity from the love of Christ.[9]

When combined with his belief that the mission of the Church was to be the sacrament of God's offer of salvation, Rahner's conviction that profane and salvation history were intertwined gives rise to two conclusions crucial to his approach to change in the Church: first, since the Church's existence in history was not for itself but for the salvation of humanity, any manifestations of ecclesial life had meaning and significance only when serving that aim;[10] secondly, although they could properly be considered citizens of the world to come, Christians did not have the right to renounce any interest in the present world, which Rahner characterized as being the world of a future which had already begun.[11] If the Christian vocation to become a sign of hope in the world was to be fulfilled, it was imperative that the Church recognize and respond to the particular contours of the present:

> Wherever and whenever one does not want to accept a situation, which is the particular situation of a particular time, but instead to flee to a world of yesterday, a dreamt-up world, a dead corner of history, a social structure which was alive and powerful yesterday, one not only fails at one's earthly task, but in such a case Christianity itself suffers both from the artificiality of this existence and the falseness of the fictitious.[12]

Inevitably, the present, to which its existence as the sacrament of God's offer of salvation committed the Church, was not static but ever-evolving. Accordingly, both during the Second World War and particularly in the decade following it, Rahner constantly reiterated the need for the Church not only to be aware of the dynamics of this changing situation, but to be courageous in meeting the challenges it presented. The remainder of this chapter will, therefore, focus on his analysis of twentieth-century Western society, his assessment of the Church's response to it, and his proposals for a more creative interaction with that world.

[9] 'History', *TI* v. 113–14 (*ST* v. 133–4).

[10] 'The Prospect for Christianity' ['Prospect'], in *Free Speech*, 87 ('Die Chancen des Christentums', in *Wort*, 61), orig. pub. 1952.

[11] 'Christianity and the "New Man"' ['New Man'], *TI* v. 149 ('Das Christentum und der "Neue Mensch"', *ST* v. 175), orig. pub. 1961.

[12] Ibid. 150–1 (ibid. 176).

THE TWENTIETH-CENTURY WORLD

Had he ventured to join the debate between the 'cyclicists' and 'linear progressivists', which has been a feature of the philosophy of history since the time of Vico, Rahner would certainly have entered the lists on the side of the latter. Since he regarded history as inextricably linked with God's offer of salvation, it followed that history could have an end only when God unveiled its deepest meanings in the Last Judgement.[13] Until then, history would remain open and in motion.[14] In Rahner's analysis, this conclusion was particularly applicable to the twentieth century.

As early as 1943, in his 'Vienna Memorandum', Rahner wrote that the last generation had been a period of extraordinary political, social, and cultural upheaval.[15] In 1952, he reiterated this conviction in characterizing the present century as a period of great transition.[16] The features of this transition were: the end of European domination of the world; the end of the isolated history of nations; and, consequently, the beginning of a period of world history, in which even the two power-blocs which emerged after the Second World War could not claim to be independent of each other.[17] Outside the political arena, Rahner regarded technology as exerting the most influence on the shape of the twentieth century.

Through technology, what had once been unthinkable had become actual: humanity had learnt to shape its own world, rather than having it shaped by nature. In the twentieth century, the chrysalis had indeed become a butterfly. As a result of modern developments, humanity had become able, for the first time in its history, to exert control over both the present and future environments.[18] While such an opinion might sound dangerously naïve in

[13] 'History', *TI* v. 99 (*ST* v. 117).

[14] 'Prospect', in *Free Speech*, 93 (*Wort*, 65).

[15] 'Vienna Memorandum', 1. This Memorandum, which was never pub., was written by Rahner while working in Vienna during the Second World War. For Rahner's own description of its background, see *I Remember*, 52–3 (*Erinnerungen*, 55–7); for further background information, see Vorgrimler, *Understanding Karl Rahner*, 68–9, and Theodor Maas-Ewerd, 'Odo Casel OSB und Karl Rahner SJ: Disput über das Wiener Memorandum "Theologische und philosophische Zeitfragen im katholischen deutschen Raum"', *Archiv für Liturgiewissenschaft*, 28 (1986), 193–234, particularly 194–9.

[16] 'Prospect', in *Free Speech*, 95 (*Wort*, 66).

[17] Ibid. (ibid.).

[18] 'Parousia', *TI* v. 310 (*ST* v. 365).

a world which, in the wake of nuclear accidents and the depletion of the ozone layer, has become hostile to uncontrolled manipulation of the environment, it should be emphasized that what was important for Rahner was not the wonder of technology *per se*, but its deeper meaning. Thus, he emphasized that the advent of the technological epoch marked a watershed in human history: the end of the era in which human beings were merely gatherers totally at the mercy of nature.[19]

Inevitably, such socio-political changes also had an impact on human values. Rahner maintained, for example, that the last three centuries in Europe had witnessed a steady increase in relativism. As a result, only the facts of sense experience, which could be confirmed by experiment, were regarded as true; all else was considered as mere theory or opinion, whose truth or falsity could not be unequivocally established.[20] In the late twentieth century, this relativism had crystallized into the conviction that, since there could be good reasons for any opinion, being a 'decent human being' was more important than what one believed.[21] Indeed, Rahner argued that contemporary society had even ceased to understand that there was a connection between what a person believed and how they acted.[22]

The evaporation of a rigorous concern for truth was not, however, the only feature of the spiritual-intellectual climate in the twentieth century which attracted Rahner's attention. Thus, in his 1949 essay, 'Change in the Form of Heresy', he suggested that never before had the variety of experiences, sciences, and knowledge available in the world been such that they could not only determine the spiritual situation of the individual, but could not themselves be mastered by any individual.[23] Although recognizing that human beings had never had exhaustive or systematic knowledge of everything that affected their lives, Rahner stressed that in the past the unmastered aspects of life related to the soil, race, talents, and other things which did not affect spiritual existence.[24]

[19] 'Prospect', in *Free Speech*, 96 (*Wort*, 67).
[20] 'Heresy', *TI* v. 470 (*ST* v. 529–30).
[21] Ibid. 471–2 (ibid. 531). [22] Ibid. 471 (ibid.).
[23] 'Heresy', *TI* v. 493 (*ST* v. 555). The 1949 essay 'Change in the Form of Heresy' appears as the third section of the 1961 article, 'What is Heresy?'.
[24] Ibid. (ibid.).

In the twentieth century, however, what remained unsynthesized were perceptions, theories, and opinions which did impinge on the spiritual level. In short, the unconquerable complexity of the age of pluralism had dawned.

Although the limited range of scientific data in the past had meant that there had been those who could claim encyclopaedic knowledge, such people were no longer to be found. Indeed, Rahner argued that nobody in a pluralist age could pretend that their level of understanding was identical with the universe of meaning.[25] While the complexity of both modern technology and contemporary institutions placed them beyond the comprehension of any individual, this complexity did not prevent them influencing the life of the individual.[26] In this way too, the twentieth century was different from previous epochs, when what people did not understand simply did not exist for them.

Even without being able to understand the processes of modern technology, people in the twentieth century none the less had to make moral choices about the products of that technology. Whereas morally and socially important decisions had once been taken in a social situation of 'almost static fixity and repetition', which facilitated the formation of explicit, universal moral norms, the contemporary context in which such decisions had to be made was qualitatively different.[27] The tempo of social change in the modern world created situations which could not be encompassed by such explicit divine ordinances as the Ten Commandments. The fact that these new moral issues—included among which, for example, was the attitude to be adopted towards nuclear weapons—outstripped the existing arsenal of didactic propositions, did not mean, however, that they were morally indifferent.[28] As will be seen, Rahner believed that the contemporary social context not only presented a challenge to the individual, but also had ramifications for the Church's pastoral practice. The impact of modern complexity on the Christian and the Church was not restricted, however, to questions of morality: it also affected the shape of belief itself.

[25] Ibid. 494 (ibid. 556). [26] Ibid. 495 (ibid. 557).

[27] 'The Church's Limits' ['Limits'], in *Christian of the Future*, 58–9. ('Grenzen der Amtskirche', *ST* vi. 506–7), orig. pub. 1964.

[28] Ibid. 56 (ibid. 505).

THE BELIEVER IN THE TWENTIETH CENTURY

In his 1949 article on heresy, to which reference has already been made, Rahner expounded the thesis that in the modern world, a society where pluralism had become insurmountable, no one could know with certainty whether they really believed. He claimed that if it had been possible to objectify the faith of an 'orthodox' person, even such faith would be found to contain some elements which were heretical.[29] What could not be determined, however, was whether the objectively heretical only constituted opinions which were not espoused with existential commitment, or, conversely, if it was the explicitly professed faith which was in fact merely a façade.[30]

To describe the unique situation of heresy in the twentieth century, Rahner introduced the wonderfully recondite term 'cryptogamic'. Since the technical application of this word is to a plant which, because it has neither stamens nor pistils, cannot bear flowers or seeds, the allusion to the essentially unproductive nature of such heresies is clear. Rahner claimed that the present era was one of widespread cryptogamic heresies, which not only remained unsystematized—a factor which thus distinguished them from what might be called the 'classical' heresies against which the Church had traditionally campaigned—but also existed alongside explicit orthodoxy.[31] Such heresies, which had first made their appearance at the time of the Modernists, did not amount to a direct attack on the Church's established doctrinal structure, but, as they exempted no one, were actually more dangerous than any previous species of heresy.[32] Nevertheless, Rahner argued that the existence of cryptogamic heresy was an inescapable aspect of life in a pluralist culture:

> We may say for the time being that everybody today is infected with the bacteria and virus of cryptogamic heresy, even though they do not on this account have to be qualified as having become diseased with it. Everyone conforms at least unconsciously and at least in the form of peripheral 'opinions' to the peripheral-existential attitudes which originate from a basic heretical attitude and in which there is sufficient *materia gravis* to constitute actual heretical attitudes.[33]

[29] 'Heresy', *TI* v. 498 (*ST* v. 560). [30] Ibid. (ibid.).
[31] Ibid. (ibid.). [32] Ibid. 499 (ibid. 561).
[33] Ibid. 500 (ibid. 563).

As with any virus, the typical feature of a cryptogamic heresy was its latency. No such heresy was expressed so explicitly that it was possible to take up a reflex position against it. Rahner argued that while such heresy avoided definite theses, its influence was often revealed by a deliberately selective presentation of faith. It could be seen, for example, in the avoidance of any reference to purgatory, hell, the evangelical counsels, or fasting; it could also be found in attitudes of mistrust or resentment against the teaching authority, attitudes which gave birth to attempts to disguise what one really thought.[34] The impact of cryptogamic heresy, suggested Rahner, made it difficult to support the contention of the 1917 Code of Canon Law that the term 'heretic' was to be applied solely to those who, after baptism, pertinaciously doubted or denied a truth of the Catholic faith.[35]

Since the influence of cryptogamic heresy excluded no one—indeed, Rahner argued that the members of the hierarchy could also be affected—it followed, according to the principles examined in dealing with Rahner's view of the sinfulness of the Church, that such heresy had to be regarded as characterizing the life of the Church itself. As in all else, the 'unconquerable grace of Christ', the expression of God's commitment to the Church, protected the Church as a whole from being so contaminated by such heresies that it could fall away from the truth.[36] As will be discussed in what follows, however, even the assistance of the Spirit, while it remained essential, did not immunize the Church against the effects of twentieth-century life.

THE CHURCH IN THE TWENTIETH CENTURY

Rahner's view of the impact of the twentieth century on the Church can be explored through his notion of 'the diaspora', an expression which became part of his vocabulary in the late 1940s.[37] In

[34] Ibid. 505–6 (ibid. 569–70).

[35] Ibid. 481. (ibid. 542). The definition of heresy in the 1917 Code can be found in canon 1325/2. Interestingly, the formula defining a heretic was not changed in the 1983 Code: see canon 751.

[36] 'Heresy', *TI* v. 502 (*ST* v. 565).

[37] Ref. to the diaspora—which the pub. trans. renders into Eng. as 'pagan surroundings'—can be found, e.g., in his 1948 article, 'Parochial Principle', *TI* ii., 288. (*ST* ii. 304). For a survey of Rahner's use of the diaspora, see Karl Neumann, 'Diasporakirche als *sacramentum mundi*: Karl Rahner und die Diskussion um Volkskirche-Gemeindekirche', *Trierer theologische Zeitschrift*, 97 (1982), esp. 52–9.

order to appreciate Rahner's understanding of 'the diaspora', it must be contrasted with 'the Christian Middle Ages'. In the latter, at least in theory, Christianity was not simply the religion of the majority, but also the principal influence on the culture of Europe. In the modern era, however, Christianity in Europe exercised no such dominance. Indeed, the opposite seemed to apply:

> we [Christians] have become strangers in the world; the world appears in its daily business to ignore us and to regard Christianity as something that has become museum material from the much-cited past of the West, something that serves the dreams and plans for world improvement of childish romantics and restoration politicians.[38]

In such a world, it could not be assumed that all people would find their home in the Church. Even faith itself could no longer be regarded as the product of natural growth within families and society, but depended on personal decision.[39] Equally, because the culture had ceased to be specifically Christian, there could be no privileged position for the Church or the clergy.[40] Furthermore, it was the State, which had both the task and the resources to develop the possibilities offered by the twentieth century, not the Church which was to be the key influence on the future.[41]

Although such changes were obviously major, Rahner did not mourn the passing of medieval Christendom. In fact, he argued that the Church's prolonged dominance over European culture had been an anomaly, which, far from reflecting the supernatural origin of Christianity or the Church, was simply the product of the closed nature of medieval society.[42] That it was not the product of a peculiar Christian genius was clear from the fact that medieval Islam and Shinto had enjoyed similar prominence in their respective cultures.[43] Rahner even asserted that, in light of the limited choices then available in technology and economics, it

[38] 'Prospect', in *Free Speech*, 61–2 (*Wort*, 43).

[39] 'A Theological Interpretation of the Position of Christians in the Modern World' ('Modern World'), in *Mission*, i. 33–4 ('Theologische Deutung der Position des Christen in der modernen Welt', in *Sendung*, 33).

[40] Ibid. 35–7 (ibid. 34–5).

[41] Ibid. 18–19 (ibid. 23–4).

[42] Ibid. 31–2 (ibid. 31–2). Rahner had orig. made this claim in 'Prospect', in *Free Speech*, 80 (*Wort*, 56).

[43] 'Modern World', in *Mission*, i. 31–2 (*Sendung* 31–2).

would have been even more of a miracle if a monoculture had not become the norm in the Middle Ages.

The Church's cultural hegemony had persisted while Europe remained closed to the rest of the world and defensive against 'outside' enemies, like the Turks or the heresies which came from the East.[44] When, however, there were no more external enemies, the challenges emerged from within. Indeed, Rahner argued that both the beginnings of the schisms within the Church and the general de-Christianization of the West through the Renaissance, Reformation, and Enlightenment—the foundation-stones of the diaspora—were contemporaneous with the period of European expansion when not only Europe, but also, ironically, the Church began to become part of the whole world.[45]

In addition to cataloguing the impact of twentieth-century life on the position of the Church in society, and on the attitudes of those who formed the Church, Rahner also assessed the Church's response to this impact. Indeed, he not only analysed how the Church had met the challenge, he even proposed an alternative response, an alternative oriented to the future. His analysis of the Church's actual response, and his proposals for an alternative response, will be considered in the next two sections.

THE RESPONSE OF THE CHURCH: THE ACTUAL

Put simply, Rahner's writings before Vatican II characterized the Church's response to the challenge of the twentieth century as woefully inadequate. In 1952, for example, he wrote that not only did contemporary Christians appear despondent and the Church itself tired, but theology and the life of the clergy showed little to encourage a positive assessment of the state of the Church.[46] Humanity's march towards a united history had aroused in many Christians only the fear of being without influence in the new world order. Indeed, Christians had become so afraid of failure, so afraid that Christianity might even disappear from the world, that they had ceased to be involved in the world.[47] As a result, the

[44] Ibid. 28–9 (ibid. 29–30). [45] Ibid. 29–30 (ibid. 30–1).

[46] For his assessment of the state of the Church in 1952, see 'Prospect', in *Free Speech*, 53–64 (*Wort*, 37–44).

[47] Ibid. 65–6 (ibid. 46).

Church often gave the impression of running behind, and in a sourly critical way, the progress of humanity:

One gets the impression that God's infinite revolution in history, in which God lets the world burn up in God's own infinite fire, rests on the shoulders of people who really put their trust in what has proved itself in the past, although this is fundamentally also only of this world and hence brittle, ambiguous and transitory, just like what belongs to the future and is still to come in this world. Why are Christians so often to be found only in the conservative parties?[48]

Conservatism and pessimism were not, however, the only negative features of the twentieth-century Church. Rahner claimed that the Church had come to resemble mass industrial societies which, since they valued only the planned and the power of the group, sought to make the individual subservient to the programmes of great institutions.[49] Such an environment, both in the Church and beyond, was particularly hostile to the charismatic.[50] Within the Church, opposition to the charismatic—an opposition more practical than theoretical—revealed itself in a one-dimensional emphasis on papal primacy as the only source of unity, and guarantee of truth, that the Church possessed. A corollary of this was the strengthening of centralist tendencies through a reliance on ecclesiastical bureaucracy.[51]

At the beginning of the 1960s, Rahner argued that the strongest yearning in the contemporary Church often seemed to be the desire not to be disturbed.[52] As a result, not only was the bureaucracy used to stifle initiative, but obedience had been reduced to a synonym for what was conducive to bureaucratic order and smoothness. Both of these features were more suggestive of a totalitarian regime than the sacrament of Christ in history.[53] The teaching office too, because it sought to extirpate the threat of cryptogamic heresy by the exercise of authority rather than creative teaching, reflected a similar passion for order as its chief priority.[54] So discouraged was Rahner by the Church's performance that he claimed the Church had, for one and a half centuries,

[48] 'New Man', *TI* v. 150 (*ST* v. 175).

[49] 'Stifle', *TI* vii. 76 (*ST* vii. 80).

[50] 'Wagnis oder Trägheit? Die Kirche und die geistige Situation der Gegenwart', *Universitas*, 18 (1963), 1209.

[51] Ibid, 1210.

[52] 'Stifle', *TI* vii. 77 (*ST* vii. 82).

[53] Ibid. 82 (ibid. 86).

[54] 'Heresy', *TI* v. 508–9 (*ST* v. 572).

failed to fulfil its mission: 'For during the modern age which is now coming to an end, its thinking and feeling, and its familiarity with the situation, have not kept pace sufficiently with modern developments; during this period it has become more of a conservative power defending itself than was right.'[55]

Rahner's vigorous criticism of the Church's response expressed his conviction that, its actual performance notwithstanding, the Church did indeed have the resources to meet the challenge of the twentieth, or any other, century. The key to any improvement in the Church's response lay not in reshaping the world—in fact, Rahner believed that the present age was no worse than any other—but in a changed attitude within the Church itself.[56] Thus, he contended that for the Church to fulfil its mission in the modern world, the great need was for courage and creativity, qualities which had been conspicuously absent in the Church's relationship with modernity.

THE RESPONSE OF THE CHURCH: THE POSSIBLE

Two fundamental principles underpinned Rahner's proposals for a more creative response to the modern world. The first was his belief that the Christian message and the Church must always relate to concrete, contemporary conditions. In his 'Vienna Memorandum', Rahner had argued that if the Church refused to develop its language, liturgy, truths of faith, religious art, and organization to respond to new conditions, it became what its enemies had always accused it of being: 'a mere aspect of a culture and a historical situation which is past or will inevitably pass away'.[57] Furthermore, he asserted that a genuine will to respect tradition implied the readiness to seek new ways and means to relate this tradition to the future.[58] In addition, this willingness to be innovative was not to be stifled for fear of making mistakes. While the possibility of false developments could not be categorically excluded, Rahner stressed that they were less to be feared than was an 'unliving traditionalism'.[59]

Rahner's second principle supporting a more energetic engagement with the present was that Christianity and defeatism were

[55] 'New Man', *TI* v. 151 (*ST* v. 177).
[56] 'Prospect', in *Free Speech*, 54 (*Wort*, 39).
[57] 'Vienna Memorandum', 2. [58] Ibid. [59] Ibid. 3.

mutually exclusive. Since the Christian was to be the one who not only hoped against hope, but walked unsupported on the water, anxiety about the future was irreconcilable with faith in God's grace guiding the Church.[60] Rahner argued that when the Church's situation was viewed even from a sociological perspective, doubts about the Church's survival could not be justified, since, as much as Islam and Buddhism in the East, Christianity was so deeply rooted in Western culture that it would not disappear without a trace.[61] He stressed, however, that this did not mean that Christians could be content merely to survive, appendix-like, in history. Nor would survival in such a form be an infallible sign of the grace of God: 'One cannot in fact quickly and easily say: if Christianity continues to exist till the end of history only in a few, as it were atavistic remaining representatives from the past, that the promises of God, that the gates of hell will not overcome the Church have been fulfilled.'[62]

If the Church was actually to flourish in the future, the first requirement was a positive assessment of the present. For Rahner, the key to this positive outlook was a recognition that the present period of history was a time of transition—the adolescent crisis of the collective consciousness—rather than of unmitigated tragedy.[63] The contemporary Church had, however, become so transfixed by a sense of loss that it faced the future with pessimism. In this, it mirrored the Church in the age of both Augustine and Gregory the Great, who had failed to realize that the period of history in which they lived was a time of growth, rather than the catastrophe they believed it to be.[64]

Even as early as the 'Vienna Memorandum', Rahner had recognized that the need to decide how much or how little to change was the source of the most profound anxiety in a period of transition. In the absence of charismatic figures, like Francis of Assisi or Ignatius of Loyola, who could indicate a new path, what was required was a tentative, careful, open-to-compromise, but still-courageous attitude towards the old and the new. While he did not deny that planning for the future would cause strains within the Church, Rahner suggested that mutual love could hold these in check.[65] In the context of changing times, the great enemies

[60] 'Prospect', in *Free Speech*, 70–1 (*Wort*, 50). [61] Ibid. 65–6 (ibid. 46).
[62] Ibid. 67 (ibid. 47). [63] Ibid. 97 (ibid. 68).
[64] Ibid. 93–4 (ibid. 65–6). [65] 'Vienna Memorandum', 4.

were a lack of courage and a lack of prudence: 'The dark and unsettling nature of such a situation must simply be borne in patience and faithful confidence and one ought not be led astray by either a reactionary or a revolutionary short-circuiting.'[66]

Rahner's refusal to underwrite the chorus of popular defeatism in regard to the position of the Church reflected not only his belief that the demise of 'the Christian West' was far from being a disaster, but also his confidence, expressed consistently in the period before Vatican II, that even the diaspora situation was not without hope. This hope came from understanding the diaspora as a 'must' in terms of salvation history.

Writing in 1954, Rahner argued that life not only could be divided into what ought or ought not to exist—such as the things prescribed or proscribed by the Ten Commandments—but also included features which, while they ideally ought not to exist, were not necessarily such a contradiction of goodness as to be without positive value for salvation. Such things were a 'must'.[67] Included among them was the existence of the Cross. Included too was the diaspora. The diaspora was a 'must' because although, in the best of all possible worlds, all people ought to be Christians, the fact that only a minority of people were actually Christians did not imply that God had forsaken either the world or the Church. Even its existence as a minority group in society did not, therefore, absolve the Church from the obligation to be a sign of contradiction in history.[68] If the diaspora was indeed a 'must', then neither the Church's retreat to the ghetto nor its abandonment of evangelization could be classified as anything other than a loss of faith.

Far from decrying the existence of the diaspora, Rahner claimed that it could facilitate the emergence of a more 'normal' Christianity than that which had characterized past 'Christian cultures'. This was so because those who lived in the diaspora would have to develop a faith which was more personal, less dependent on the institution and the traditional, and less conditioned by its surroundings.[69] In the same vein, the 'Vienna Memorandum' had

[66] Ibid. 3.

[67] 'Modern World', in *Mission*, i. 20–4 (*Sendung*, 24–6). The same idea is also discussed in 'Prospect', in *Free Speech*, 81 (*Wort*, 56–7).

[68] 'Modern World', in *Mission*, i. 25–7 (*Sendung*, 27–8).

[69] 'Unbelieving Relations', *TI* iii. 359 (*ST* iii. 423).

stressed that in a period of social upheaval, the Christian was challenged to focus more consciously on the central reality of Christianity: the graced relationship to the trinitarian God through the mediation of Christ—this concentration on essentials was to become a primary focus of Rahner's response to pluralism.[70] The Christianity of the diaspora would, therefore, be significantly different from medieval or baroque Christianity, which could scarcely be distinguished from the prevailing culture. In the diaspora, there could be no 'folk-costume Christianity', which relied on social norms rather than personal commitment.[71] The Church of the diaspora would be dependent on the good will of its members, rather than on its institutional strength.[72]

The diaspora also made it imperative that the Church reconsider the possibilities and limits of its contribution to the world. Accordingly, in an article written after the beginning of Vatican II, but before the appearance of *Gaudium et Spes*, Rahner rejected the notion that the Church could effortlessly provide concrete prescriptions conducive to universal peace and happiness.[73] Indeed, he claimed that the gap between the universal moral principles proclaimed by the Church and the concrete prescriptions which individuals and societies used to shape their lives was more pronounced in the twentieth century than ever before.[74] Since the Church did not have a greater insight into the future than the average person, it could not assert that it alone could lead humanity to that future. Rahner's refusal to regard the Church as *sedes omniae sapientiae* was consistent both with his conviction that no single interpretation of any issue could claim proprietorial rights over the designation 'Christian', and also with his belief that there was no specifically Christian plan for the State, culture, or economics.[75] That there seemed to have been in the past was only because other possibilities had been limited.

That the Church was no longer the sole arbiter of social values did not imply, however, that the outlook for Christians in the world was irredeemably bleak. While acknowledging the possibility

[70] 'Vienna Memorandum', 25.
[71] 'Unbelieving Relations', *TI* iii. 359 (*ST* iii. 423).
[72] 'Modern World', in *Mission*, i. 36 (*Sendung*, 34–5).
[73] 'Limits', in *Christian of the Future*, 50 (*ST* vi. 500).
[74] Ibid. 67–8 (ibid. 513–14).
[75] 'Modern World', in *Mission*, i. 13 (*Sendung*, 19).

that, in the area of morality, the Christian in the diaspora might indeed be brought into conflict with the values of society, Rahner stressed that the world ought not to be portrayed as prone to moral decay simply because the Church and the clergy were not in charge.[76] In fact, he maintained that if the Christian were to suffer under the contradictions which could exist between faith and the world, such an experience of martyrdom might actually be the beginning of authentic discipleship.[77]

The Church's existence in history meant neither that it needed to direct the whole of that history nor even that its office-bearers ought to devise a concrete plan for the intramundane future, a plan based on principles which only Christians could advocate.[78] Rahner not only strongly defended the autonomy of the world, but also emphasized that a world not administered by clerics was not necessarily a world without a sense of God. Indeed, he argued that efforts by the Church to exercise control over everything would ultimately be damaging for the Church itself: 'The insistence for the sake of the ghetto, on integrating everything into an ecclesial framework naturally means that the clergy have to be in control of everything. This results in an anti-clerical feeling which is not always an effect of malice and hatred for God.'[79]

Although he accepted that materialism and technology seemed to have submerged the religious instinct in modern society, Rahner remained convinced that this instinct could not be suppressed indefinitely.[80] Furthermore, he was confident that Christianity had much to offer a world entering a new period of its history; indeed, he assembled a battery of arguments to highlight what Christianity could offer. First, he suggested that, since a unified world-history would need a religion which was neither regionally nor culturally limited, the universalism of Christianity was greatly to be valued. Secondly, he emphasized that the idea of the oneness of humanity, which a world history incarnated, was a distinctly Christian insight.[81] Thirdly, Christianity's ability to address itself to the dark sides of life, such as suffering and death, could enrich

[76] Ibid. 35 (ibid. 34).
[77] 'Unbelieving Relations', *TI* iii. 360 (*ST* iii. 424).
[78] 'New Man', *TI* v. 151 (*ST* v. 176).
[79] 'Modern World', in *Mission*, i. 43 (*Sendung*, 39).
[80] 'Prospect', in *Free Speech*, 98 (*Wort*, 69–70).
[81] 'New Man', *TI* v. 152 (*ST* v. 178).

society, since even the technology of the twenty-first or twenty-second centuries would not be able to eliminate the causes of human suffering.[82] Finally, he believed that Christianity would appeal as its historical existence and tradition were unique in a world accustomed only to the immediate products of human ingenuity.[83]

Successful evangelization in the diaspora required, however, that the Church avoid duplicating the error that it had often made in mission lands: trying to impose a culture as well as the Gospel.[84] Thus, even while accepting the value of structures such as the parish, Rahner stressed that the Church's methods of pastoral care needed to respond to the life-style of those who heard the Gospel, rather than trying to tailor everyone to existing structures. More important than a particular ecclesiastical culture was that people were coming to faith: 'One soul apostolically won from a milieu which has already reverted to paganism is worth three hung onto from the remnants of traditional Christianity (one would almost like to say folk-costume Christianity).'[85]

The challenge for the Church's hierarchy in the modern world was to present the Church's message and life in a way which did not create any unnecessary difficulties for those who might actually be open to that message and that life. Rahner argued that successfully meeting this challenge required: a reform of the liturgy; a new demonstration of Christian morality so that it did not appear to be incomprehensible imperatives imposed from without, but the expression of what was objectively right; and the establishment of a relationship between clergy and laity which, while conserving the permanent structure of the Church, was more than patriarchism.[86] In short, as he had suggested in the 'Vienna Memorandum', Rahner believed that the well-being of the Church depended on its presenting itself as something other than the *societas perfecta*:

Only when Christians see in the Church today more than a confessional organisation; only when they also see the visible Church as the medium through which they stand in a most profound, grace-filled relationship with Christ, and therefore with divine life; only therefore when the

[82] 'Prospect', in *Free Speech*, 103 (*Wort*, 73). [83] Ibid. 104 (ibid.).
[84] 'Modern World', in *Mission*, i. 48–9 (*Sendung*, 42–3).
[85] Ibid. 52–3 (ibid. 45). [86] 'New Man', *TI* v. 151–2 (*ST* v. 177–8).

doctrine of the mystical body of Christ—whether in connection with this title or without it, is irrelevant—is a more alive and personal possession for them, will they remain faithful to the Church, even in this age.[87]

Over and above the specific areas where he wanted to see reforms, Rahner stressed the need for both the clergy and the laity to concern themselves with plans for the future. For him, it was not enough simply to restate eternal principles; also required was the courage to make concrete suggestions to meet particular needs. Such suggestions could be made without implying that they embodied uniquely Christian insights.[88] It was also important, however, that Church members be patient with the Church as it sought to respond to the world. To expect immediate answers to contemporary questions was to ignore the fact that answers, no less than the questions themselves, proceeded from a history of reflection.[89]

Rahner's proposals for a more positive response to the modern world were characterized by a spirit of hope. He did not predict a short-term improvement in the Church's prospects in society, much less a new 'Christian era', but he evinced no doubt that the Church had a role to play in shaping the future. He stressed that engaging the world was part of the vocation of the Church in every age. Unless the Gospel bore the imprint of each age, it could be a sign of the Church's indifference to the truth, a sign of the heresy of 'dead orthodoxy'.[90] Consequently, the Gospel must have the imprint of the twentieth century, an imprint which could be had only if the Church acted with courage.

At the heart of Rahner's conviction that the Church had nothing to fear from the present was his belief that contemporary trends such as individual autonomy and a demythologized view of the world actually had their roots in Christianity:

If we consider the doctrine of freedom and of absolute responsibility for self—and the doctrine stating that the particular fate (and eternity) of each individual person is the result of their own free acts—then it becomes clear that the possibility gradually dawning on people today, viz. the possibility of making themselves the object of their planning and

[87] 'Vienna Memorandum', 32.
[88] 'Modern World', in *Mission*, i. 15 (*Sendung*, 20–1).
[89] 'Limits', in *Christian of the Future*, 66 (*ST* vi. 512).
[90] 'Heresy', *TI* v. 507–8 (*ST* v. 571).

formation, is merely the echo and particular application of that deeper self-responsibility which Christianity has always acknowledged the person to possess and which it has always steadfastly refused to relieve them of, since it has always regarded it as their own—sometimes painful—burden.[91]

In place of the fear that had characterized the Church's response to the modern world, Rahner urged the need for a prophetic voice. This voice was to affirm the legitimacy of human planning, but also remind humanity that its future was to be found in Christ, rather than its own achievements.[92] In regard to its own members, the proper task of the Church was not to remove the complexity of the twentieth century, but to strengthen believers in their pilgrimage of faith through a rapidly evolving history:

> the decisive task of the Church is definitely not a facile preparation of a concrete model for such a Christian life, a model which one would only have to copy obediently, honestly and comfortably in order clearly to be a good Christian. For that life, the Church does not offer a model, but the strength to endure that life without models and it offers this strength precisely through fulfilling its proper religious tasks.[93]

Rahner's hope that the Church might learn to respond more prophetically to the modern world was not purchased at the cost of abandoning his criticism of how the Church had actually performed. Indeed, so convinced was he of the inadequacy of the Church's response that, as late as 1962, he regarded the Church as being in danger of extinguishing the Spirit.[94] It was from such a perspective that he assessed the prospects of Vatican II.

VATICAN II: POSSIBILITIES AND PROSPECTS

In an essay published in 1962, Rahner outlined his expectations of Vatican II. His opinions in that article were characterized by a sober reserve, perhaps even pessimism, rather than an unbridled enthusiasm. The foundation of this restraint was his conviction that the effectiveness of the Council would necessarily be impaired

[91] 'New Man', *TI* v. 153 (*ST* v. 179).
[92] 'Parousia', *TI* vi. 311–12 (*ST* vi. 366–7).
[93] 'Limits', in *Christian of the Future*, 72 (*ST* vi. 518).
[94] 'Stifle', *TI* vii. 79 (*ST* vii. 83–4).

by the fact that the Spirit in the Church had been partially extinguished.[95] Thus, he claimed that the weakness of contemporary theology made it unlikely that the Council would be able to communicate the Christian faith in a way that addressed the needs of the present. Not even a Council, he argued, was capable of overcoming the obstacle posed by an impoverished theology:

> one cannot expect any doctrinal decrees which will make the non-Christian sit up and take notice to any particular extent and which will fill the minds and hearts of Christians with a completely new and unusual light. To demand this would be quite out of keeping with the nature of a Council in present-day circumstances. The magisterium cannot replace the charism of theology and it is not its job to do so. If this charism is weak today, then this will manifest itself also in the doctrinal decrees of a present-day Council.[96]

On the basis of his negative assessment of pre-conciliar theology, Rahner specifically cautioned the Council against a rash of doctrinal decrees. While such decrees were valuable when a Council met specifically to resolve a disputed question, he stressed that such was not the background of Vatican II. The real threats to the Church in the twentieth century and beyond were those born out of positivism, cryptogamic heresies, and pluralism.[97] The challenge presented by these issues could not be determined by dogmatic pronouncements, but, as was also indicated above, only by proclaiming and living the Christian faith with renewed courage.

Where Rahner was prepared to express confidence that the Council could achieve positive results was in the realm of Church discipline. He identified topics such as decentralization—within the limits created by the need to respect both the Church's unity and the responsibility of each part of the Church—greater openness to oriental and Protestant churches, the simplification of canon law, reform of the liturgy in light of the impetus provided by the liturgical movement, restoration of the diaconate, and renewal of the laws of fasting, as areas where the Council could provide valuable directions.[98] In a comment rich in prophetic insight, Rahner noted that although decisions on such matters might

[95] 'On the Theology of the Council', *TI* v. 259 ('Zur Theologie des Konzils', *ST* v. 294).

[96] Ibid. 262 (ibid. 298).

[97] Ibid. 263 (ibid. 298–9).

[98] Ibid. 264–5 (ibid. 300).

at first sight appear to be very harmless, self-evident and not very far-reaching solutions [they can] in reality have an as yet incalculable effect on the future and on the outlook of people within the Church, an effect which perhaps cannot even be estimated by the very authors of any such decrees concerned with Church discipline, the liturgy, educational discipline or pastoral matters.[99]

As an example of the possible effects of such decisions, he argued that a decree on the oriental churches could ultimately serve as a model for the future status of the Church in Africa or Asia, areas which, like the East, had their own cultural character.[100] Underpinning this example was his conviction that such Churches could not remain forever subsumed under the Latin model.

As in his view of the relationship between the various charismatic elements in the Church, Rahner stressed that no Council decrees, whatever their wisdom, could substitute for the work of the Spirit within the Church. Thus, he claimed that even a decree on the use of Scripture in the liturgy, theology, and Christian life would not necessarily produce either a love for the Scriptures or a 'biblical movement'.[101] Ultimately, therefore, it was the Spirit alone who would make the work of the Council bear fruit for the future well-being of the Church:

Even after the Council, the Church will still be the Church of sinners, of pilgrims and of those who must search laboriously, the Church will always obscure God's light again by the shadows cast by her children. All this is no reason for not having a Council or for expecting nothing or too little from it. Here too God's power will become mighty in our weakness. Much may be decided which God will then change in God's own way into grace and blessing for the Church and humanity. Each person and the Church must do what belongs to them. They must sow and plant in patience! It is wonderful that even in the Church and for the Church all increase nevertheless depends on God and that we can hope for this without any merit on our part.[102]

REVIEW

In dealing with the Church's existence in history, Rahner was guided by one general principle: 'The Church cannot choose the

[99] 'On the Theology of the Council', *TI* v. 265 ('Zur Theologie des Konzils', *ST* v. 300–1).
[100] Ibid. (ibid. 301). [101] Ibid. (ibid.). [102] Ibid. 266–7 (ibid. 302).

situation in which it lives. The situation is given to the Church and therefore it is good.'[103] Fully reconcilable with this principle are the two convictions which were reflected in his analysis of the Church's situation in the twentieth century: first, his refusal to be pessimistic about the change from the 'Christian West' to the diaspora; secondly, his commitment both to identifying the challenges which the modern era posed for the Church and to formulating creative responses to those challenges, responses which called for changes in the Church's law, structures, and practices.

In the earlier chapters of this work, the emphasis was on the God-given dynamism which Rahner saw as intrinsic to the Church's beliefs and structures. In this chapter, however, his perception of the modern world has been highlighted as the single biggest influence on his advocacy of the Church's need to change. What began to appear in Rahner's work in the 1940s and 1950s was the conviction that the effectiveness of the Church in contemporary society, its ability to preach and embody the Word of God in ways the denizens of the twentieth century would not find alienating, depended on its willingness to make substantial changes both to its internal operations and to its relationship with the world. What was needed was a new openness, a willingness to compromise, and a willingness to take risks in making changes, even when it could not be known with certainty whether such changes were fully reconcilable with tradition.

Inevitably, the broad sweep of Rahner's approach to the need for change in the twentieth-century Church left several significant 'loose ends'. Thus, while emphasizing that the demise of the bond between Church and society would create the need for a strong personal faith if faith was to survive in the diaspora, he did not address the related issue of how this faith would connect with the Church and its structures. A similar lack of detail characterized his encouragement for pastoral strategies designed to meet particular needs. While this opened the way for a myriad of approaches to pastoral care, Rahner did not specify how unity was to be maintained between these differing strategies. What is clear, however, is that even before Pope John XXIII committed the Church to *aggiornamento,* Rahner had committed himself to a changing Church. The following chapter will concentrate on the shape of this commitment after Vatican II.

[103] 'Christian Living Formerly and Today' ['Christian Living'], *TI* vii. 4 ('Frömmigkeit früher und heute', *ST* vii. 12), orig. pub. 1966.

5

'Changing Forward': The Impact of Vatican II

In terms usually applied only to political scandals or scientific discoveries, the blurb for one English-language edition of the documents of Vatican II referred to the Council as 'world-shaking in its impact'.[1] While this was an assessment of the content of the Council's documents, it could be argued that both John XXIII's surprise announcement of his intention to call a Council and its proceedings—which included the rejection of the initial schemata prepared by the pontifical commissions, the unofficial lobbying of national blocs of bishops, and the numerous tense moments on the floor of the Council itself—could be described in similarly dramatic terms.[2] Thus, Vatican II was not simply another event in the history of the Church, but was the single most important influence on the shape of contemporary Catholicism.

[1] The edn. referred to is Austin Flannery (ed.), *The Documents of Vatican II* (New York, 1975).

[2] Amongst the plethora of books and articles dealing with the background and conduct of the Council, an informative contemporary account (with numerous refs. to Rahner) is Xavier Rynne, *Vatican Council II* (New York, 1968); a later, briefer, overview can be found in John W. O'Malley, *Tradition and Transition: Historical Perspectives on Vatican II* (Wilmington, Del., 1989), 11–18; various reminiscences of those involved in the Council are given in Alberic Stacpoole (ed.), *Vatican II Revisited by Those Who Were There* (Minneapolis, 1986)—this collection includes Herbert Vorgrimler's 'Karl Rahner: The Theologian's Contribution', 32–46—and in Klinger and Wittstadt (eds.), *Glaube im Prozeß*, see esp. 16–182; for Rahner's involvement before and during the Council, see Karl-Heinz Neufeld, 'Theologen und Konzil: Karl Rahners Beitrag zum Vatikanischen Konzil', *Stimmen der Zeit*, 202 (1984), 156–66; Rahner's response to the various pre-conciliar schemata can be found in the opinions he wrote for Cardinal König, see Herbert Vorgrimler (ed.), *Sehnsucht nach dem geheimnisvollen Gott* (Freiburg i.B., 1990), 95–165; excerpts from letters Rahner wrote while in Rome for the Council are given in Vorgrimler's, *Understanding Karl Rahner*, 141–84; Rahner's own assessment of his contribution to the Council can be found in *I Remember*, 81–90 (*Erinnerungen*, 89–100).

Not surprisingly, Rahner was quick to realize the potential of the Council for reshaping the Church, In this chapter, the emphasis will be on his assessment of the ramifications of Vatican II for the life and mission of the Church, and specifically on the Council as an agent of change.

Despite the fact that he co-edited a volume of Council documents, Rahner's activity in the years after Vatican II was not limited to exegesis of the various conciliar texts.[3] In fact, his main concern was to further the renewal of the Church which Vatican II had inaugurated. Far from regarding Vatican II as the final chapter in the history of the Church, Rahner contended that the mission of proclaiming the Gospel in an ever-evolving world could be fulfilled only if it was accepted that the Council had neither spoken the last word about the shape of Christian living, nor left theology with nothing to discuss. To support his summons to ongoing renewal, Rahner claimed both that such change was fundamental to the nature of the Church, and that Vatican II itself had expressed the principles which ought to guide renewal. The arguments he advanced to substantiate such claims will be examined in what follows.

THE CHURCH: CHANGEABLE AND UNCHANGEABLE

Shortly before the end of the Council, Rahner addressed himself to the pain felt by those who already regarded Vatican II as rupturing the Church's continuity. While acknowledging that fidelity to the Word of God and respect for tradition, rather than surrender to the demands of ephemeral fashions, were features of a Catholicism which could properly be esteemed as 'conservative', he also stressed that that same tradition included freely initiated changes in law and doctrine.[4] To appreciate how he could maintain that both conservation and change were legitimate features of Church life, what is required is an understanding of his distinction between divine law and positive ecclesiastical law.

Rahner argued that divine law, an instance of which was the indissolubility of a valid Christian marriage, since it expressed

[3] Together with Herbert Vorgrimler, Rahner edited: *Kleines Konzilskompendium* (Freiburg i.B., 1966).

[4] 'The Changing Church', in *Christian of the Future*, 11. (*ST* vi. 456–7), orig. pub. 1965.

God's will for the whole Christian era, was unchangeable—where 'change' suggests repeal or substantial amendment. Conversely, ecclesiastical law, such as the ban on cremations, since it represented the Church's response to particular socio-cultural situations, could be altered in accord with the exigencies of history.[5] Thus, already present in that early response to the Council was the principle that remained a hallmark of Rahner's post-conciliar theology—and which, therefore, will be of major importance in the following chapters: the Church had both the responsibility to retain immutable divine law and the right to change ecclesiastical law.

While Rahner thus quashed the charge that the Church's refusal to abrogate divine law indicated a lack of either generosity or understanding of the needs of a particular age, he also emphasized that, within the parameters set by immutable divine law, the Church had both the prerogative and obligation to make changeable regulations for the 'spiritual good' of its members.[6] Since the Church was composed of diverse groups with diverse needs, Rahner accepted that debates on the speed and opportuneness of change in ecclesiastical laws were inevitable; what was beyond question, however, was the fundamental legitimacy of change. Indeed, he claimed that a refusal to change could imply that the Church was unfaithful to its Lord:

> anyone who appreciates the rapid change in historical circumstances and does not flee from this into a ghetto; anyone who knows that there is, and has always been, a mutable human law in the Church, and that this kind of change has always been practised . . . will recognise in the change itself the immutable which would be betrayed by rigid immutability: fidelity to the eternal Gospel and obedience to the Lord of history, which both go to produce the change in the Church's legislation.[7]

As outlined by Rahner, the phenomenon of change in the Church could not be limited solely to questions of discipline, but also embraced dogma. Indeed, in regard to dogma, he was committed to what might initially seem to be two contradictory claims: on the one hand, he maintained that when the Church's teaching office required the unconditional assent of faith for a particular doctrine, such a doctrine was not revisable; on the other, he also

[5] 'The Changing Church', in *Christian of the Future*, 13–15. (*ST* vi. 458–60), orig. pub. 1965.

[6] Ibid. 16 (ibid. 460). [7] Ibid. 20 (ibid. 464).

stressed that even unchangeable dogma had a history, not merely before, but also after, its formal definition.[8] Reconciling these two views requires an examination of Rahner's understanding of 'change' as it applied to dogma.

CHANGING THE UNCHANGEABLE DOGMA

As has already been noted, Rahner's theory of the development of dogma, with its emphasis on a change in, not of, identity, stressed that change and continuity were reconcilable. Building on this foundation, his post-conciliar work argued that to claim a history for dogma even after its definition suggested not rejection, but a deepening of the faith of the Church which that dogma embodied—a 'changing forward', not a 'changing backwards'.[9] Understood in this way, change implied a growth in understanding, a clarification of the connection with other doctrines, a new formulation for a new age, or the highlighting of features conducive to ecumenism. The impetus for this change came from the Church's existence in history.

Since each period of history not only had its own concerns, but raised questions which no previous age could have anticipated, all defined dogmas, because they were the product of a previous age, could be called into question. The answer to such questions was an enduring truth which could itself be expressed only in changing contexts:

> realisations of truth always remain open to further modifications in the future. The history of the Church's faith is still in process of development, and every statement of faith posited in the here and now, and truly posited at that, can be modified and transformed by factors belonging to a future that is still unknown.[10]

Indeed, Rahner suggested that statements of faith not open to development were, in fact, a 'dead chapter of fixed formulas', rather than signs pointing to the living God.[11] Far from being merely the substitution of new truths for old, change was, therefore,

[8] Ibid. 22–3 (ibid. 465–6). [9] Ibid. 24 (ibid. 467).

[10] 'The New Image of the Church' ['New Image'], *TI* x. 5 ('Das neue Bild der Kirche', *ST* viii. 331), orig. pub. 1966.

[11] *Grace in Freedom* [*Grace*], trans. H. Graef (London, 1969), 10 (*Gnade als Freiheit* [*Gnade*] (Freiburg i.B., 1968), 200).

a means of preserving truth. Since the truth by which the Church lived was not a series of propositions, but God's own life, this truth could never be exhausted by human formulations. The fulfilment of the Church's belief would not, therefore, be found in a sacrosanct statement about God, but only in the vision of God: 'Until then even the enduring, permanently valid truth is only partial, spoken in images and parables, wandering and therefore changing on the pilgrim road of unpredictable history.'[12]

In the years before Vatican II, Rahner's reflections on the development of doctrine had focused on defined dogma; after the Council, he turned for the first time to the Church's panoply of non-defined propositions. The justification for these non-defined teachings he located in their practical application. Thus, he argued that, just as life in any other sphere did not simply hover between the poles of belief and non-belief but was shaped by things about which there could not be absolute certainty, the Church's doctrine, concerned as it was to express the unfathomable mystery of God, did not consist merely of what must either be believed with absolute assent or be freely rejected.[13] Non-defined propositions, like those things the individual believed even though they had not been totally proved, remained valid and necessary until a new, more complete, insight emerged. For this reason, such propositions were entitled to the respect and assent of the believer.

Non-defined propositions, particularly in moral theology, were often the Church's response to conditions which, like nuclear weapons or the possibility of artificial birth-control, had not previously existed. Rahner argued that while the Church in such cases could not draw ultimate answers from the Gospel, the fact that such issues were not, as was noted in the previous chapter, morally indifferent, also meant that it could neither allow them to remain unaddressed by Christian reflection nor simply leave believers to their own resources.[14]

When faced with such new situations, the Church attempted to provide guide-lines by applying basic principles. While acknowledging that the wisdom of particular non-defined teachings could be obscured by the Church's over-hasty decisions, short-sightedness,

[12] 'Changing Church', in *Christian of the Future*, 25 (*ST* vi. 468).
[13] Ibid. 26–7 (ibid. 468–9). [14] Ibid. 28 (ibid. 470).

or lack of understanding—all of which belonged to the Church's servant form and pilgrim nature—Rahner nevertheless judged it to be necessary that the Church respond to contemporary issues.[15] Indeed, he suggested that even those teachings which did not claim absolute assent could, because they involved the least danger of coming into conflict with the unchangeable Spirit of the Gospel, be binding for the faith and moral conscience of the believer. The person accepting the Church's non-defined teachings was, therefore, like the patient guided by a doctor's informed, but not wholly certain, opinion that an operation was necessary.[16]

His commitment to the value of the Church's non-defined doctrine notwithstanding, Rahner also affirmed the believer's right not to follow such reformable teaching. While he was far from advocating frivolous opposition to the Church, Rahner maintained that rejection of a particular teaching could be justified for those who, while self-critically aware of their own capacity for short-sightedness, had examined their conscience before God and were prepared to defend their decision before God's judgement-seat.[17] Rahner characterized the mature Christian as one who could distinguish between the unchangeable dogma of the Church and that which could always be better expressed and more nuanced.[18] Thus, while promoting respect for the Church's teaching authority, he was also anxious to assert the need for, and legitimacy of, ongoing development of reformable teaching—a theme which will be prominent in the next chapter. Since Rahner's analysis of the dynamics of change in the Church was undertaken in the light of what had occurred at Vatican II, the next task is to explore why he regarded the Council as an example of change.

THEOLOGY IN A NEW KEY

In rejecting the accusation that Vatican II had impaired the Church's inherited faith, Rahner stressed that not only had the Council reverenced the *depositum fidei*, but that not even the most progressive of the Fathers had suggested that any defined dogmas ought to be disregarded.[19] There was, therefore, never the slightest

[15] Ibid. 29 (ibid. 471). [16] Ibid. 30 (ibid. 472).
[17] Ibid. 31 (ibid.). [18] *Grace*, 11 (*Gnade*, 201).
[19] *Konzilskompendium*, 19–20.

possibility that even controversial dogmas, such as those on papal infallibility or the Immaculate Conception and Assumption, might be revoked.[20] Nevertheless, Rahner was in no doubt that, without obscuring received truth, Vatican II represented a legitimate development of dogma. This development had taken place in such issues as: how the papal primacy and the episcopate could work together; how the necessity of the Church for salvation could be understood as compatible with the possibility of salvation for those outside the Church; and deeper insight into Mary's place in the economy of salvation.[21]

Despite his affirmation that it was unreservedly committed to continuity, Rahner did acknowledge that, if compared with the theology of the nineteenth century, Vatican II could indeed seem more revolutionary than developmental. That such an impression was possible, however, reflected the weaknesses of pre-conciliar Scholasticism rather than any anarchic tendencies of the Council. Indeed, Rahner characterized the century before Vatican II, the period which embraced the reigns of the Popes from Pius IX to Pius XII, as the age of 'Pian monolithism', an essential feature of which was its tendency to regard all important questions of theology as having reached their ultimate determination.[22] The conventional wisdom of the time was that once a dogma had been defined, there could be no further development, no 'changing forward' in the sense outlined above. In the twentieth century, this approach left room for progress only in such peripheral areas as Mariology—which had become a growth industry in the years preceding the Council.[23] Vatican II broke the Pian mould.

Neither the theological method nor the language of the Council simply echoed neo-Scholasticism. While the Council Fathers

[20] 'Changing Church', in *Christian of the Future*, 22 (*ST* vi. 466).

[21] Ibid. 24 (ibid. 467).

[22] 'The Second Vatican Council's Challenge to Theology' ['Challenge'], *TI* ix. 6 ('Die Herausforderung der Theologie durch das Zweite Vatikanische Konzil', *ST* viii 16), orig. pub. 1966. Unfortunately, the English version translates *Pianische Monolithismus* only as 'monolithism'. Rahner's definition of the Pian period can be found in *I Remember*, 87 (*Erinnerungen*, 97); see also his 'Rückblick auf das Konzil', in *Toleranz in der Kirche* [*Toleranz*] (Freiburg i.B., 1977), 107–10.

[23] 'Challenge', *TI* ix. 5 (*ST* viii. 16). A detailed study of the features of 19th-cent. theology is Gerald A. McCool's *Catholic Theology in the Nineteenth Century: The Quest for a Unitary Method* (New York, 1977)—in the light of Rahner's criticism that pre-conciliar theology was part of a monolithic structure, the book's subtitle is significant.

recorded their conviction that the theology of Thomas remained valuable, they did not bind themselves to Thomism.[24] In another departure from the approach of the Pian era, Vatican II's awareness of the complexity of many questions, its desire that such questions be carefully studied, and its commitment to enter into dialogue with those whose knowledge derived from sources other than Christian revelation, meant that the Council had refrained from claiming to have the answers for all disputed points of theology.[25] One concrete result of this restraint was that Vatican II did not proceed with the definition of Monogenism as dogma, even though such a proposal had been contained in one of the schemata.[26]

Rahner regarded the theology of the Council as representing a transitional stage. Passing into oblivion was a neo-Scholasticism which not only used the New Testament for proof texts, but which concerned itself with theologumena such as limbo; succeeding it was a more biblically oriented theology which not only eschewed its predecessor's 'theological exuberance' in Mariology, but was more ecumenical in approach.[27] What became clear at the Council was that theology could say something important without needing to elevate everything to dogma. The contributions which this new style of theology made to the Council were extensive. Among them were: highlighting the sacramentality of episcopal ordination; the doctrine that the whole episcopate, with and under the Pope, was the supreme ecclesiastical authority; and the acknowledgement that the human authors of Scripture were genuine authors.[28]

It was not, however, only in its approach to doctrine and theology that the Council initiated significant changes. In fact, Rahner regarded its achievements in these spheres as minor when compared to its impact on the history of the Church, an impact which

[24] The ref. to Thomas can be found in sect. 10 of *Gravissimum Educationis*.

[25] 'Challenge', *TI* ix. 5 (*ST* viii. 15–16).

[26] 'On the Situation of the Catholic Intellectual' ['Catholic Intellectual'], *TI* viii. 98 ('Zur Situation des katholischen Intellektuellen', *ST* vii. 372), orig. pub. 1966. That Rahner was particularly anxious that Monogenism not be defined is clear from his response to its inclusion in the schemata: see Vorgrimler (ed.), *Sehnsucht*, 101–3.

[27] 'The Abiding Significance of the Second Vatican Council' ['Abiding Significance'], *TI* xx. 94 ('Die bleibende Bedeutung des II Vatikanischen Konzils', *ST* xiv. 308), orig. pub. 1979.

[28] Ibid. 95 (ibid. 309).

he described as launching a new era in that history. The grounds for this assertion, and its implications for the future of the Church, will be examined in what follows.

A NEW ERA OF CHURCH HISTORY

As was noted at the end of the previous chapter, Rahner's writings immediately prior to the Council were generally pessimistic when dealing with the prospects of Vatican II. Yet, in 1979, with the benefit of much reflection on the Council, he advanced the thesis that if the history of the Church was understood not in politico-cultural terms, but as the history of the nature of the Church, then Vatican II actually represented a 'caesura', a transition to a new stage, in that history.[29] As such, the Council ranked in significance with the only other events which Rahner accepted as caesuras: the launching of the mission to the gentiles, and the linking of Christianity with Hellenistic-European civilization.[30] In regard to Vatican II, what needs to be investigated is how Rahner understood this caesura as coming about, its precise meaning, and the implications of it for the post-conciliar Church.

Rahner asserted that the changes which constitute a caesura were not usually the product of prior theological reflection, but rather resulted from a 'secret instinct' of the Spirit and the working of grace.[31] While conscious reflection was neither to be despised nor to be regarded as superfluous, it could not claim primacy over the charismatic dimension. Consequently, neither the fact that Vatican II had focused on the nature of the Church nor the fact that it was a 'Council of the Church about the Church' could be portrayed as the result of the conscious decision of any one person or group.[32] If it had indeed come about through a 'secret instinct' of the Spirit, then the Spirit could be said to work through history, since the issues which acquired importance in the life of the Church were decided by developments in that history.[33] Rahner argued that in addition to the unplanned, and, indeed,

[29] 'Basic Theological Interpretation of the Second Vatican Council' ['Theological Interpretation'], *TI* xx. 84 ('Theologische Grundinterpretation des II Vatikanischen Konzils', *ST* xiv. 296). Rahner's outline of how Church history is to be understood can be found on p. 78 (p. 288).

[30] Ibid. 83–4 (ibid. 294–5).

[31] Ibid. 85 (ibid. 297).

[32] 'New Image', *TI* x. 4–5 (*ST* viii. 329–30).

[33] Ibid. 5 (ibid. 331).

unforeseeable, prominence which ecclesiology had acquired at the Council, what also qualified Vatican II to be ranked as the third Spirit-engineered caesura in the annals of the Church was its commitment to the world-Church.

In referring to the world-Church, Rahner was not simply implying that Vatican II marked the beginning of the Church's presence outside Europe. In fact, as was indicated in the previous chapter, he recognized that, as a result of its links to European expansionism and colonialism, the potential for the Church to become a world-Church had existed for four centuries. During that period, however, the Church had actually reflected an 'export firm' dispatching an essentially European product to the four corners of the earth, rather than a Church at home in strikingly different cultures.[34] What made Vatican II different was that the bishops from the 'mission lands', far from simply being European expatriates, genuinely represented the Church in their homelands. Consequently, the Council had been

> a first assembly of the world-episcopate, not acting as an advisory body to the Pope, but with him and under him as itself the supreme teaching and decision-making authority in the Church. There really was a world-council with a world-episcopate such as had not hitherto existed with its own autonomous function.[35]

Despite his enthusiasm for the notion of the world-Church, Rahner did not naïvely suppose that the post-Vatican II Church would never again know a dominant European influence. Nevertheless, he was insistent that the movement towards a world-Church, however rudimentary and timid it might have been at the Council, represented the wave of the future. Recognition of the fact that the Church of the future would not simply be composed of the whole world, but would also have to proclaim the Gospel within a world which was not culturally homogeneous, was reflected in the Council's endorsement of vernacular languages, its affirmation of God's universal salvific will, and also in its conscious adoption of an ecumenical perspective.[36]

As has been indicated above, Rahner valued the notion of the world-Church for its capacity to provide direction for the future of the Church. In fact, it was this concern for the future which

[34] 'Theological Interpretation', *TI* xx. 78 (*ST* xiv. 288).
[35] Ibid. 80 (ibid. 290). [36] Ibid. 81–2 (ibid. 292–3).

shaped his reading of all the Council's documents. The following sections will identify what he regarded as the most significant aspects of Vatican II for those seeking to construct a theology for the future of the Church.

SACRAMENTALITY AND SALVATION

Although there were aspects of *Lumen Gentium*—Vatican II's constitution on the Church—which he found inadequate, Rahner was none the less committed to the basic 'model' of the Church which the document propagated. That this was so is hardly surprising when it is remembered that the Council endorsed the understanding of the Church which was central to Rahner's own pre-conciliar ecclesiology: the Church as the sacrament of salvation for the world (*LG* 1, 9, 48). The Council's adoption of the sacramental approach did not, however, resolve all ecclesiological questions; indeed, it inevitably gave birth to new questions.

Principal among the issues raised but not resolved by the constitution on the Church was how to reconcile the doctrine that membership of the Church was necessary for salvation (*LG* 14) with the declaration that all those who act in good faith, the unbaptized included, could in fact attain to salvation (*LG* 16).[37] While Rahner's theology of the 'anonymous Christian' offered a way out of this dilemma, this option did not find its way into the constitution. None the less, he argued that the acknowledgement of both God's universal salvific will—a notion endorsed not only in *Lumen Gentium*, but also in *Ad Gentes Divinitus* (*AG* 3) and *Gaudium et Spes* (*GS* 2)—and of the salvific value of non-Christian religions—to be found in *Nostra Aetate* (*NA* 2)—was to be included among the major insights of Vatican II. Indeed, he was convinced that such a development had contributed to the caesura which the Council represented. For Rahner, the dramatic nature of such a caesura not only revealed that the Spirit had been at work in the Council, but also indicated that it would take time to find solutions to the problems left unresolved by the Council:

> no one who maintains in principle the radical difference between truth and error, who has recognised a true claim to absoluteness on the part of Christianity and the Church, who grants in principle to certain formulated

[37] 'New Image', *TI* x. 12 (*ST* viii. 338).

insights and religious institutions a significance that has a part in deciding humanity's eternal destiny, can regard the caesura which came in with the Council as something to be taken for granted. Such a person must recognise it as a fundamentally Christian event, as a victory of Christianity and not of liberalism. Such a person must be prepared to put up with and work on all the theological problems involved in such a change—something that is not at all easy and will remain a task for a long time.[38]

While not suggesting that a solution to the problem of the Church's relationship to the salvation of non-Christians could be easily devised, Rahner was none the less convinced of the parameters within which any proposed explanation ought to lie. These were: rejection of the idea that Christians were the elect who alone were on the way to salvation; the need to affirm the Christian belief that the Church was the sole source of blessedness; and, the necessary consequence of such a belief, the avoidance of an 'ecclesiological relativism' which suggested doctrinal issues were secondary to the need to find a Church which suited one's tastes.[39]

Consistent with what he had taught before the Council, Rahner's writings after Vatican II continued to emphasize that the baptized were not the sole travellers on the roads leading to salvation. They were rather 'the uniformed section of God's campaigners'—the number and identity of those who campaigned without such a 'uniform' being known to God alone.[40] Indeed, while not disputing that God alone could know whether the offer of grace had been accepted or rejected, Rahner was more willing after the Council than before it to be positive about the prospects of salvation for those outside the Church. Thus, via a rich metaphor, he encouraged those concerned about the ultimate future of the non-believer to trust that God's saving will ensured that: 'the morning light on the mountains is the beginning of the day in the valleys, not the day above condemning the night below'.[41]

The conviction that God's salvific will was at work outside the Church did not, however, lead Rahner down the path of the 'ecclesiological relativism' he specifically rejected. The Church

[38] 'Abiding Significance', *TI* xx. 99 (*ST* xiv. 314).

[39] 'New Image', *TI* x. 13 (*ST* viii. 338–9).

[40] 'The Teaching of Vatican II on the Church and the Future Reality of Christian Life' ['Future Reality'], in *Christian of the Future*, 84 ('Konziliare Lehre der Kirche und künftige Wirklichkeit Christlichen Lebens', *ST* vi. 484), orig. pub. 1965.

[41] Ibid. (ibid. 485).

was not to be understood in juxtaposition to the grace offered to the unbaptized, but rather as the embodiment of that grace in history.[42] Thus, even in the *Handbuch der Pastoraltheologie*, where he argued that the success of the Church's missionary efforts depended on adapting its strategies and structures to contemporary conditions, he continued to define the Church in terms resonant with those he had used before the Council: 'The Church is the community, legitimately constituted in a social structure, in which through faith, hope and love, God's eschatologically definitive revelation (God's self-communication) in Christ remains present for the world as reality and truth.'[43]

Thus, what remained central to Rahner's thought was the conviction that the Church was the sacramental sign of the grace offered to the world and history as a whole. Furthermore, since the Spirit of Christ was what constituted the Church, it was possible to claim not only that Christ was present in the Church, but also that the Church was the presence of Christ in the world. At the same time, however, Rahner discouraged any reference to the Church as 'Christ living on'.[44]

The fact that the grace of Christ both founded the Church and, through it, offered salvation to all, meant that the Church could properly be portrayed as not only the fruit of salvation, but also the means by which this salvation was offered to the whole of humanity.[45] The Church was not, therefore, just another institution, not just another element of the twentieth century's social-intellectual pluralism, but was rather:

> the apprehensibility of that which already unifies at the interior level, as the historical expression of that which is universal. . . . the sheer rendering present of the God-planned nature of humanity as subject to God's designs . . . as the *fundamental sacrament* of a grace which, precisely because

[42] 'New Image', *TI* x. 14 (*ST* viii. 340).

[43] *Theology of Pastoral Action*, trans. W. J. O'Hara (Freiburg i.B., 1968), 26–7 (*Handbuch der Pastoraltheologie i* (Freiburg i.B., 1964), 118–19; the *Handbuch* was co-authored with Franz X. Arnold, V. Schurr, and L. Weber). For a survey of Rahner's ecclesiology in the *Handbuch*, see H. Schuster, 'Karl Rahners Ansatz einer existentialen Ekklesiologie', in Vorgrimler (ed.), *Wagnis Theologie*, 370–86.

[44] 'On the Presence of Christ in the Diaspora Community according to the Teaching of the Second Vatican Council' ['Diaspora Community'], *TI* x. 92–3 ('Über die Gegenwart Christi in der Diasporagemeinde nach der Lehre des Zweiten Vatikanischen Konzils', *ST* viii. 416–17), orig. pub. 1966.

[45] Ibid. 93 (ibid. 417).

it is offered to all, presses forward to express its sacramental historical nature even where the individual sacrament (of baptism) has not yet been conferred.[46]

Rahner argued that if the Church was the sacrament of salvation for the world, then it functioned as a leaven within the mass which was humanity. This meant that its influence could not be confined to where it was seen to take effect, but was actually at work everywhere and for all.[47] In this, the Church was like any other sacramental sign whose presence in history was the guarantee that the grace of God was operative even where it could not be seen. The graced offer of salvation, which was always made through the Church, was not contingent on the physical presence of the Church.[48] Thus, prior to any contact with the Church, even the unbaptized could have responded to Christ's invitation to follow him. Consequently, there could be no justification for Christians to regard themselves as the élite community of salvation. Rather than succumbing to élitism, Catholic Christians needed to recognize both that their membership of the Church was an unmerited grace, and that God alone knew whether their life as Christians was characterized by that faith and love which *Lumen Gentium* (*LG* 14) had identified as constituting the fullness of membership.[49]

Properly understood, sacramental ecclesiology not only liberated Christians from the arrogance of exclusivity, but, conversely, also meant that they need have no anxiety about being a minority in the world. Indeed, Rahner asserted that the very insignificance of the Church in the late twentieth-century world was commensurate with the Church's nature as a sacrament: 'The sign of the mystery of light in darkness can only be modest and almost insignificant. The message of what is to come (and that is what the Church is) cannot itself be what is to come; the Church of time is not as vast as the eternal kingdom of God.'[50]

This did not mean, however, that Rahner counselled the Church to abandon its missionary efforts. Indeed, in the *Handbuch der Pastoraltheologie*, he made clear that missionary activity was integral to the nature of the Church; significantly, however, he stressed

[46] 'New Image', *TI* x. 23 (*ST* viii. 348). Rahner's emphasis.
[47] Ibid. (ibid. 348–9). [48] Ibid. 14 (ibid. 339–40).
[49] Ibid. 17 (ibid. 342).
[50] 'Future Reality', in *Christian of the Future*, 93 (*ST* vi. 491).

missionary activity ought to be undertaken in a 'dialogical' manner which, because it respected local cultures, distanced it from colonialism.[51] In addition, missionary efforts were to be guided by the realization that baptism did not denote the negation of what preceded it—which would have implied that the unbaptized were totally lacking in grace—but involved the Church calling people to a conscious recognition of the God who was already at work in their lives. Consequently, the Church's message needed to be: 'become what you are'.[52]

The Church which would be entered by those who responded to the invitation to be baptized was indeed the sacrament of Christ, but was not a Church freed from weakness. What the Council itself taught about the flawed nature of the Church, and Rahner's assessment of its teaching, will be discussed in what follows.

THE SINFUL CHURCH

In Chapter 1, Rahner's analysis, from 1947, of the Church's holiness and sinfulness was discussed. In that early study, he argued that the Church was holy because of the indwelling of the sanctifying Spirit, and sinful because its existence as a visible reality, rather than as a mystical sodality, meant that it was affected by the sinfulness of its members. Both in his pre-conciliar assessment and in his post-Vatican II reflections on the Church's sinfulness, Rahner claimed that any failure to acknowledge such sinfulness, any portrayal of the Church as independent of the flawed Christians who found their home within it, misrepresented the Church's true nature.[53] The strength of his conviction on this issue dictated his response to the Council's discussion of sinfulness and the Church. Thus, while he was encouraged by the Council's references to the theme, he was disappointed at the lack of thoroughness with which it was pursued.

The allusions in *Lumen Gentium* to the Church's sinfulness—'allusions' only as the discussion of the Church's sinfulness did not merit even a complete paragraph—Rahner found wanting in

[51] *Handbuch der Pastoraltheologie*, ii/2 (Freiburg i.B., 1966), 61–2.

[52] 'Future Reality', in *Christian of the Future*, 93 (*ST* vi. 491).

[53] 'The Sinful Church in the Decrees of Vatican II' ['Sinful Church'], *TI* vi. 277 ('Sündige Kirche nach den Dekreten des Zweiten Vatikanischen Konzils', *ST* vi. 328), orig. pub. 1965.

clarity, intensity, and detail.[54] His basic criticism was that the tenor of the document envisaged a Church moving from good to better, rather than from sin to pardoning grace.[55] Despite this weakness, he recognized in the Council's stress on 'the pilgrim Church' the potential for a deeper awareness of the sinful nature of that Church. This was so because acknowledgment of the Church's pilgrim status implied that the Church was neither a heavenly entity untouched by history nor an 'institute of salvation' which cared for people while not identifying itself with them.[56] Similarly, Rahner suggested that the resonances of 'the people of God' as a synonym for the Church could, more obviously than allusion to 'the mystical body', accommodate a sinful Church.[57]

Rahner claimed that calls for renewal of the Church—a popular conciliar motif best represented by the reference in *Lumen Gentium* to the Church as *semper purificanda* (*LG* 8)—had meaning only if it was first accepted that the Church was indeed the subject of sin and guilt. However, neither *Unitatis Redintegratio* (*UR* 4) nor *Lumen Gentium* (*LG* 8 and 9), two of the documents voicing such a call, were prepared to identify what it was from which the Church needed to be purified. Rahner was also clearly disappointed that, despite the general acknowledgement that the Church was afflicted by the sins of its members, no document actually employed the term 'the sinful Church'.[58]

His insistence that the theme of the Church's sinfulness ought be given a more prominent place did not mean, however, that Rahner had abandoned his sacramental ecclesiology, with its emphasis on the Church as the means of grace in the world. Indeed, he did not resile from his conviction that the holiness of the Church, effected as it was by God's eschatological action, had priority over its sinfulness. In addition, he stressed that, even antecedent to the actual behaviour of its members, the Church was protected by this grace against falling away from God in a fundamental way:

> God's making available by predestination efficacious grace for the Church as a whole has already overtaken, without abolishing or damaging, the freedom of humanity and it has been revealed in the victory of Christ; the

[54] Ibid. 279 (ibid. 331).
[55] Ibid. 280 (ibid.).
[56] Ibid. 282 (ibid. 334).
[57] Ibid. (ibid. 333).
[58] Ibid. 284–5 (ibid. 336–7).

predicate of holiness is for the Church, which has been brought into being by God and not by humanity, the most real and decisive as opposed to its sinfulness.[59]

What remained consistent, therefore, in the twenty years between Rahner's two reflections on the sinful nature of the Church was his insistence that the Church, far from being a hypostasis, was holy only because of the sanctifying action of God. The efficacy of this grace was not to be seen in the Church becoming immune to sin, but rather in the fact that individuals and the Church as a whole sought God's mercy to free them from their sinful state.[60] What was crucial for Rahner was that the Church not forget that it was actually a pilgrim whose perfection would come only with the consummation of the Kingdom:

> We are always playing the incomplete symphony of the glory of God and it is always only a dress rehearsal. But all the hardship, the always incomplete and not able to be completed reform is not in vain, not senseless. It is simply the task of servants who sow in tears, so that God may harvest . . .[61]

Clearly, Rahner would have preferred Vatican II to specify that this 'incomplete symphony' of Christian life found its home in a Church whose sinfulness meant that it too was incomplete. Nevertheless, he was more satisfied—if still not completely content—with the Council's teaching that the primary venue for such a performance was to be 'the local Church'.

THE LOCAL CHURCH

Without great fanfare, and within a chapter concerned with the hierarchical ministry of the Church, *Lumen Gentium* (*LG* 26) referred to the Church in each local area as making present the Church of Christ. While, as was pointed out in Chapter 2, Rahner had already made use of this idea in the years before the Council, it assumed an even more prominent status in his ecclesiology in

[59] 'The Sinful Church in the Decrees of Vatican II' ['Sinful Church'], *TI* vi. 291 ('Sündige Kirche nach den Dekreten des Zweiten Vatikanischen Konzils', *ST* vi. 343–4), orig. pub. 1965.

[60] Ibid. 292 (ibid. 345).

[61] 'Was wurde erreicht?', in Rahner, Oscar Cullmann, and Heinrich Fries (eds.), *Sind die Erwartungen erfüllt?* (Munich, 1966), 31.

the period after Vatican II. Indeed, he was convinced that the constitution on the Church could have actually begun with the theme of 'local Church', as this expressed the 'highest truth' applicable to the Church as a whole:

> in it [the local Church] Christ himself, his Gospel, his love and the unity of believers are present. The Constitution recognises and explicitly states that the local and altar community, so far from being a mere minor administrative subdivision in a major religious organisation called the Church, is actually the concrete reality of the Church, the presence of Christ in which it achieves its highest fullness, and that too in the word, in the Eucharistic meal and (even in evangelical theology this point may perhaps not always be expressed with the clarity that could be desired) in the love which unites the hearers and those who celebrate the Eucharist.[62]

Crucial for Rahner was the awareness that if a local community actually made present 'the Church', rather than merely a segment of it, then the Church could not properly be understood if viewed solely as the *societas perfecta* in which everything of value originated from the top.[63] Furthermore, if it was in the celebration of the Eucharist, the sacrament of Christ's presence, that the Church was most fully actualized, then the diversity in liturgy which had been encouraged by the Council's support for vernacular languages would mean that local Churches could develop their own character rather than simply being 'administrative districts of a totally and homogenously organized state'.[64] While the local community envisaged by *Lumen Gentium* was a diocese united under its bishop, Rahner argued that the document also opened the way for the development of a theological, rather than merely canonical, approach to the parish, since it too was an altar community.[65] Turning to the future, he claimed that it would be from where people first encountered the Gospel and the Eucharist—namely, the local community—that their fundamental understanding of the Church would derive. Such Christians would not interpret

[62] 'New Image', *TI* x. 10 (*ST* viii. 335). The pub. trans. refers to 'parish community', rather than 'altar community'.

[63] Ibid. (ibid. 336).

[64] 'Abiding Significance', *TI* xx. 92 (*ST* xiv. 305).

[65] 'Pastoral-Theological Observations on Episcopacy in the Teaching of Vatican II' ['Episcopacy'], *TI* vi. 367 ('Pastoraltheologische Bermerkungen über den Episkopat in der Lehre des II Vatikanum', *ST* vi. 429–30), orig. pub. 1965.

their experience as merely happening in the Church, but rather as the realization of the Church.[66]

His emphasis on the local community did not imply, however, that Rahner envisaged the Church becoming so fragmented that, in the future, 'churches' would replace 'the Church'. In fact, he stressed that each local community would need to be aware of the importance of being united with all the other communities. Only by their union in the one Church could each community be assured of remaining in the truth; only the hierarchically constituted, and divinely ordained, union of the communities of Christ could properly express the Church.[67] Rahner argued that this union of the manifold local communities would be maintained by each community applying, within the 'hierarchy of truths', the principles of *Lumen Gentium*.[68] He did not, however, attempt to specify the mechanics of this unity. Instead, he argued—in a way consistent with the principles discussed in Chapter 3—that the Church's unity, together with the proper ordering of the relationship between the episcopate and primacy, or clergy and laity, could be accomplished only by the Spirit. Since it was an aspect of the mystery of the Church, a reflection of the fact that Christ was one and the Spirit was one, unity could not be guaranteed solely by human reason.[69]

This Spirit-generated unity of the various local Churches Rahner regarded not as a 'secondary amalgam' of communities which originally existed independently of one another, but as the Church which was always one.[70] The community gathered around the altar was an expression of the concrete reality of the Church as a whole, rather than an alternative to it. While the dynamics of the relationship between the local and universal Church remained to be specified, Rahner was none the less certain of one thing concerning the future of the local communities: that their home would be in the diaspora.

THE CHURCH OF THE DIASPORA

Although *Lumen Gentium* referred to the diaspora solely as a possible location in which the local Church might be found (*LG* 26),

[66] 'New Image', *TI* x. 11 (*ST* viii. 336–7).
[67] Ibid. (ibid. 337). [68] Ibid. (ibid.).
[69] 'Diaspora Community', *TI* x. 88, n. 7 (*ST* viii. 412, n. 7).
[70] Ibid. 100 (ibid. 423).

Rahner postulated that its acknowledgment of the diaspora's existence would prove to be the constitution's most important legacy to the Catholic of the future.[71] Indeed, while not attempting to predict whether it would take twenty or even one hundred years for the conditions associated with the diaspora to evolve, he was convinced that the future of the Church would be experienced within a context where Christians were 'a little flock'.[72] As discussed in the previous chapter, Rahner was firmly of the opinion that neither a homogeneous Christian culture nor 'the Catholic nation' was to be resuscitated—'Everywhere will be the diaspora, and the diaspora will be everywhere.'[73]

Furthermore, he argued that its existence in the diaspora would reshape the Church. As a minority group lacking political influence or social prestige, the Church of the future would not, for example, be able to rely on the eminence of its institutions. Instead, if it was to convince non-members of the virtue of its views, it would have to rely on 'the strength of the heart'.[74] In addition, since it would emphasize the priority of the local community, the diaspora would also radicalize the understanding of what it meant to be a member of the Church. No longer would there be merely nominal members, but those identifying themselves as Christians would be those who

> gathered around the altar, announcing the death of the Lord and entrusting the darkness of their own lot—a darkness which no one will be spared even in the super-welfare state of the future—to the darkness of the death of their Lord. They will know that they are like brothers and sisters to one another, because there will be few of them any more who have not by their own deliberate decision staked their own heart and life on Jesus the Christ; there will be fewer fellow-travellers, for there will be no earthly advantage in being a Christian. They will certainly preserve faithfully and unconditionally the structure of their sacred, unworldly community of faith, hope and love, the Church as it is called, as Christ founded it.[75]

As was indicated in the previous chapter, Rahner believed that the diaspora was not simply a situation which Christians were obliged to tolerate because there were no alternatives, but one that actually facilitated the manifestation of the Church's true nature. The very insignificance of the Church in the society of the

[71] 'Future Reality', in *Christian of the Future*, 78 (*ST* vi. 480).
[72] Ibid. (ibid.). [73] Ibid. 79 (ibid.). [74] Ibid. 80 (ibid. 481–2).
[75] Ibid. (ibid. 481). The pub. trans. omits the refs. to 'fellow-travellers'.

future, its sharing in the crucifixion of Christ, was to be its strength. So convinced was Rahner of the theological value of a community witnessing to insignificance and poverty, that he claimed that the diaspora communities provided the model for what the Church should always be.[76] Thus, only in the diaspora could it truly be grasped that, for example, the love expressed in the Eucharist was neither that of Christ's unveiled glory nor of 'idealistic philanthropy', but that of the darkness of death.[77]

Rahner's understanding of the diaspora also accorded with his sacramental view of the Church. Since the Church itself was not the Kingdom of God made manifest in glory, but rather the sign and promise of that Kingdom, the community hoping against hope for its fulfilment, it was appropriate that the Church should be contradicted by the world, just as its Lord had been so rejected.[78] Its existence as the sacrament of Christ also meant that the Church in the diaspora would have the task of finding a proper balance between pride and humility. Although the Church could be proud of being the sign of the salvation offered in Christ, Rahner was clear that the diaspora also called for the Church to be humble before the fact that it was only the sign, rather than the realization, of that salvation.[79]

While he believed that even the world-Church of the future would be a Church existing in the diaspora, Rahner nevertheless proscribed the notion that Christians could therefore be justified in retreating to their 'upper room' and fearfully closing out the world. Indeed, he was particularly blunt in warning the Church of the future against becoming a sect. Such a sect—'a group of individuals who are inadequate to meet the demands of life'; '[people] who glorify a ghetto-like way of life as lived in the protected places of history'—could not fulfil the Church's mission to involve itself actively in shaping the world of the future.[80] Rahner was not, however, content merely to caution against sectarianism, he also advanced a range of proposals for a more creative relationship between the Church and the world.

As has already been noted, Rahner was convinced that the application of genuinely Christian principles, not an abject surrender to secular pressure, was the proper inspiration for change

[76] 'Diaspora Community', *TI* x. 96 (*ST* viii. 420).
[77] Ibid. 97 (ibid. 420–1). [78] Ibid. (ibid. 421).
[79] Ibid. 98 (ibid. 422). [80] Ibid. (ibid. 421–2).

in the Church. In Vatican II's proposals for altering the Church's relationship to the world, he found a shining example of the application of such principles. Accordingly, he lauded both *Gaudium et Spes* and *Dignitatis Humanae* for renouncing external means in matters of religious conversion, for promising respect even to erroneous consciences, and for affirming the legitimacy of a properly secular world outside of the Church's control.[81] Similarly, he praised the Council's general abjuration of a 'clerical-fascist' worldview—his pejorative description of the idea that the Church, understood as the clergy, needed to be in charge of everything.[82]

The Council's spurning of any suggestion that the Church ought to exercise hegemony over the world was not, however, a disguised option for that sectarianism so abhorrent to Rahner. In fact, he interpreted the Council as a clear statement by the Church that it was not prepared to languish in the backwaters of history. In stark contrast to that anguished desire to be left alone which was the breeding ground of sectarianism, the fact that the Council committed the Church to the service of humanity implied a willingness to participate in the developing history of the world.[83] This commitment to service was eloquently expressed in the preface to *Gaudium et Spes*, which proclaimed that nothing which affected humanity could remain alien to the followers of Christ (*GS* 1). Indeed, Rahner regarded the whole of *Gaudium et Spes*, the Council's final document, as the work of a Church seeking neither to dominate nor flee the world, but to realize its place in it: 'In *Gaudium et Spes*, in an act of the whole Church as such, the Church as a whole became expressly aware of its responsibility for the future history of humanity.'[84]

Rahner was convinced that, if it was to play a role in the unfolding history of the world, the Church needed both to learn about that world and to engage in dialogue with it. Learning about the world was necessary because the twentieth-century world was not simply the immediate product of God's creation, but had also been shaped by human dreams and decisions, the impact of history, and the reflections of sociology and other scientific disciplines.[85] Similarly, dialogue was required in order to understand

[81] 'Abiding Significance', *TI* xx. 92–3 (*ST* xiv. 306).
[82] Ibid. 93 (ibid. 307). [83] *Grace*, 13 (*Gnade*, 203).
[84] 'Theological Interpretation', *TI* xx. 81 (*ST* xiv. 291).
[85] 'Challenge to Theology', *TI* ix. 7 (*ST* viii. 17–18).

those whose view of the world was not influenced by Christian revelation. In Rahner's opinion, this dialogue could not simply be a new subdivision of theology, but had to become part of all aspects of theology, since no theological disciplines could claim immunity from contemporary developments in the human sciences.[86] That Rahner himself did more than acknowledge the need for knowledge of, and relationship with, the modern world, is clear both from his work on the *Handbuch der Pastoraltheologie*, which will be discussed in more detail in the next chapter, and his colloquies with groups as diverse as Marxists and natural scientists.[87]

It was in the Church's relationship to the scientific community that Rahner believed that the Council itself had provided a model of healthy development. Thus, he praised *Dei Verbum* for accepting—even if only cautiously—the application of modern historical-critical and philological methods to the biblical sciences (*DV* 22), and *Gaudium et Spes* (*GS* 52) for recognizing the contribution which the secular sciences could make towards understanding and enriching marriage and family life.[88] Rahner interpreted Vatican II's openness to such developments as laying to rest the ghost of Galileo. Furthermore, he was convinced that neither scientists nor dogmatic theologians needed to be anxious that the truths of their respective disciplines would contradict each other.[89]

Rahner not only believed that the Church had nothing to fear from the contemporary rational-critical spirit, but also argued that through its dialogue with the world, the Church could actually be enriched in its own inner life—as an example, he suggested that scientific method could teach the Church about the importance of development. Just as scientific method required an openness to new processes, experiments, and results, so Rahner claimed that it was no cause for regret that the Church, in its non-defined teachings, was also obliged to remain open to a possible need for more nuanced statements.[90] While continuing to affirm the value, and even the binding nature, of such teaching, he also stressed

[86] 'Challenge to Theology', *TI* ix. 7 (*ST* viii. 17).

[87] See e.g. his comprehensive analysis—some of which was written with Norbert Greinacher—of the modern world in ch. 7 of *Handbuch der Pastoraltheologie*, ii/1 (Freiburg i.B., 1966), 178–266. For details of the various dialogues in which Rahner was involved, see Vorgrimler, *Understanding Karl Rahner*, 111–16.

[88] 'Catholic Intellectual', *TI* viii. 98–9 (*ST* vii. 372–3).

[89] Ibid. 101 (ibid. 375).

[90] Ibid. 101–2 (ibid. 375).

that a conversation with science would provide the Church with an opportunity to clarify both the content and level of obligation attached to its non-defined teaching.[91]

While the Church had much to learn about, and from, the modern world, Rahner stressed that the dialogue between science and the Church was not simply to be a case of the former influencing the latter. Accordingly, just as modern science challenged the Church to a more careful formulation of its teachings, the Church's faith in a transcendent mystery challenged scientists to see that human beings could not be understood by an exclusive concentration on their rationality.[92] Science and its 'eggheads' needed, therefore, to be reminded that they did not constitute 'the authoritative magisterium for the ultimate interpretation of being'.[93]

Significantly, Rahner emphasized that the Church's commitment to dialogue did not involve the abandonment of its claim to be an 'absolute society'. Accordingly, he stressed that the Church was not a debating circle, where everything and anything could be said, doubted, and done in a radically different way from what had been the norm.[94] Thus, Vatican II could not be interpreted to mean that the Church no longer saw itself as mediating God's definitive self-communication in Christ. The model of how such an exclusive claim could be reconciled with openness to an evolving world Rahner found in *Gaudium et Spes*.

For Rahner, the importance of the Council's pastoral constitution lay in its nature as an 'instruction'. This instruction he defined as a counsel to a concrete decision in a particular situation, a decision which could not simply be derived from a universal principle.[95] The Church's right to issue such instructions was grounded in its mission and power to guide the Christian actions of the individual. Where a direct application of a universal principle was not possible, the Church sought to guide the individual by attempting to discern, 'according to the best of its knowledge and conscience', what was most conducive to the realization of Christian life in a specific situation.[96]

[91] Ibid. 103–4 (ibid. 377). [92] Ibid. 107 (ibid. 379).
[93] *Handbuch der Pastoraltheologie*, ii/2. 276.
[94] 'Catholic Intellectual', *TI* viii. 100 (*ST* vii. 373–4).
[95] 'On the Theological Problems Entailed in a "Pastoral Constitution"', *TI* x. 298 ('Zur theologischen Problematik einer "Pastoralkonstitution"', *ST* viii. 618).
[96] Ibid. 303–4 (ibid. 623–4).

Both the possibility and value of such instructions derived from the Spirit. The Spirit it was which empowered a response to a particular situation; the Spirit it was which gave binding force to the Church's instructions.[97] The crucial role of the Spirit did not, however, exclude the fact that the recognition of what was needed in a particular situation was a human recognition which could not be derived solely from revealed truth.[98] Consequently, the Church was obliged to draw on sources of knowledge which it did not control. Hence the need for dialogue. Although during the Council itself there had been opposition to the Church's reliance on non-revealed knowledge, Rahner stressed that the work of the Spirit in the Church was not irreconcilable with scientific methods of analysis.[99]

Rahner's analysis of Vatican II's impact on the Church's relationship with the world was complemented by his exploration of the Council's implications for the shape of life within the Church itself. In this context, his view that the Church of the diaspora would not only be a Church without worldly significance, but also a Church where believers were bound together more by active faith than by all-embracing structures has already been noted. What this situation meant for the future of the episcopal ministry will be the subject of the next section.

THE EPISCOPATE: ITS MINISTRY AND ITS FUTURE

One conclusion which can be drawn from Rahner's analysis of the Council is that Vatican II had been a sterling example of what episcopal ministry could contribute to the Church. If there were those who believed that the Council Fathers had been isolated from the real concerns of the Church, this criticism found no echo in Rahner. In fact, he claimed that the results of the Council would not have been significantly different even if lay participants had been included.[100] Rahner's post-conciliar reflections on the episcopacy did not, therefore, embrace the possibility of its abolition. The two issues which did absorb his energy in regard to

[97] 'On the Theological Problems Entailed in a "Pastoral Constitution"', *TI* x. 304–6 ('Zur theologischen Problematik einer "Pastoralkonstitution"', *ST* viii. 624–6).
[98] Ibid. 308 (ibid. 627–8). [99] Ibid. 311 (ibid. 631).
[100] *Konzilskompendium*, 23–4.

bishops were: the proper ordering of episcopal ministry and papal primacy—the question which he had addressed even before the Council; and the role of the bishop in the diaspora community.

As was discussed in Chapter 2, seeking an alternative to the model which regarded bishops as merely the Pope's subalterns had been of major interest for Rahner in the years immediately prior to the Council. This same theme was developed in *Lumen Gentium* (*LG* 18–23).[101] The endorsement of episcopal 'collegiality' which can be found in those passages meant that the constitution echoed what had been Rahner's own conclusion: with the Pope as primate, the college of bishops formed the highest authority in the Church.[102] As interpreted by Rahner, Vatican II's identification of the united episcopate as the upholder of supreme power in the Church proved that debates over whether such power properly belonged to the Pope or a Council were ultimately vacuous.[103] While the Council had endorsed the importance of collegiality, it did not attempt to detail the dynamics of the college of bishops. Rahner, however, was not so reticent.

The episcopal college, he argued, existed in 'two modes': either the whole college could act together or the Pope, as its head, could act alone.[104] When the Pope did act alone, it was an 'act of the college'—because the Pope, the supreme member of that college, expressed the college in his actions—but not a 'collegiate act'—which occurred only when the whole college, with the Pope, was involved.[105] While the action of the Pope was the action of the college, the Pope did not require the college's approval before being able to act:

In the strictest and most precise sense, the Pope has a supreme power not as a figure over and against the College or above it, for this college is in fact itself indisputably vested with this same supreme power. The position is, rather, that the Pope exercises this same power as an individual also, and this is precisely why in his possession of this power he is set

[101] Rahner's detailed commentary on *Lumen Gentium*, secs. 18–27, can be found in Herbert Vorgrimler (ed.), *Commentary on the Documents of Vatican II*, i, trans. K. Smyth (London, 1967), 188–207.

[102] 'Episcopacy', *TI* vi. 361–2 (*ST* vi. 423–4).

[103] 'On the Relationship between the Pope and the College of Bishops', *TI* x. 52 ('Zum Verhältnis zwischen Papst und Bischofskollegium', *ST* viii. 376), orig. pub. 1967.

[104] Ibid. 55 (ibid. 379).

[105] Ibid. 58 (ibid. 382).

apart from every other individual bishop and also from the sum total of all the bishops.[106]

Rahner stressed that the Pope and the college were not originally two separate powers which had been combined into one, but had always existed as a unity. Since 'supreme power' could not be divided, the Pope and the college implied each other.[107] On the vexed question of how this worked in practice, Rahner reiterated his long-held conviction that harmony could be achieved only through the Spirit, rather than canonically. At the same time, however, guided by the principle that an authority which is not exercised is no authority at all, he advocated positive steps to give bishops a more expanded role in the Church's supreme government.[108] Such steps, which will be specified below, he interpreted not as diluting power in the Church, but as making concrete the real nature of that power.

While anxious that the relationship between the Pope and the bishops be seen to be more collegial, Rahner did not countenance either a permanent Council or the suggestion that certain bishops reside in Rome to act as lobbyists on behalf of the world episcopate. What he did favour was greater episcopal representation in the congregations of the Curia, in order that what emanated from Rome might be more attuned to the whole Church.[109] His other concrete proposal was for an 'episcopal advisory board', which would involve representatives of bishops' conferences meeting regularly with the Pope and being superior to the Curia.[110]

The years after the Council did not, however, see Rahner's dreams realized. Thus, writing in 1979, he expressed his disappointment that the breakthrough represented both by the affirmation of collegiality in the Council's documents and by the processes of the Council itself had not been consolidated in such post-conciliar developments as the Synod of bishops.[111] Despite this disappointment, he urged the continuation of efforts to develop a balanced relationship between the Pope and the bishops. For Rahner,

[106] 'On the Relationship between the Pope and the College of Bishops', *TI* x. 58–9 ('Zum Verhältnis zwischen Papst und Bischofskollegium', *ST* viii. 376), orig. pub. 1967.

[107] Ibid. 63 (ibid. 386). See also *Theology of Pastoral Action*, 82–4 (*Handbuch der Pastoraltheologie*, i. 168–9).

[108] 'Relationship between Pope and Bishops', *TI* x. 70 (*ST* viii. 393–4).

[109] 'Episcopacy', *TI* vi. 365 (*ST* vi. 427).

[110] Ibid. (ibid. 427–8).

[111] 'Theological Interpretation', *TI* xx. 80 (*ST* xiv. 291).

collegiality, not a return to Pian centralism, represented the best hope of meeting the needs of the world-Church.[112]

Rahner's interest in collegiality was essentially an interest in the local Church. Collegiality meant that the local bishop needed to take a greater responsibility for the development of the Church in his own diocese, rather than relying solely on directives from Rome. Reiterating what had appeared in his pre-conciliar writings, Rahner stressed that the health of the universal Church depended on developments in each local Church.[113] It was the local bishop, not a Roman functionary, who was best able to appreciate the needs of his diocese. It was, therefore, also the local bishop, the chief pastor of the diocese, who ought to respond to those needs.[114]

While it might seem that Rahner's advocacy of the rights of the local bishop was an extravagant interpretation of Vatican II's approach to collegiality, Rahner believed that it was the Council itself which had opened the way for a greater exercise of initiative at the local level. As support for this conviction, he could point to the fact that *Lumen Gentium* (*LG* 29) had directed that the episcopal conference of each nation, rather than Rome, was the proper body to decide whether or not a restoration of the permanent diaconate was opportune.[115]

Turning to the future, Rahner argued that a symbiosis would exist between the place of the bishop in the Church and the primacy of the local Church. In other words, only if the bishop was clearly seen to be associated with the local Church would the role of the episcopal ministry be valued:

If the empirically real structure of the Church in the local community is really present and capable of being experienced, the episcopal nature of the Church's essential structure, which lies as it were at a deeper level in the organism of the Church, will become accessible to the religious experience of the Christian and will no longer give the impression of being

[112] Ibid. 89 (ibid. 302). The Eng. trans. refers to 'the centralism which was usual in the time of Pius XII'. This does not, however, capture the full import of *die pianische Epoche*, which, as has been noted, includes the time from Pius IX until Pius XII.

[113] 'Episcopacy', *TI* vi. 363 (*ST* vi. 425).

[114] 'Theology and the Church's Teaching Authority after the Council' ['Teaching Authority'], *TI* ix. 91 ('Kirchliches Lehramt und Theologie nach dem Konzil', *ST* viii. 121), orig. pub. 1966.

[115] 'The Teaching of the Second Vatican Council on the Diaconate', *TI* x. 227–8 ('Die Lehre des Zweiten Vatikanischen Konzils über den Diakonat', *ST* viii. 546–7), orig. pub. 1966.

an abstract theory which has little contact with the concrete life of the Church.[116]

What was also to be true of bishops in the Church of the future, the Church of the diaspora, was that they would not have any earthly honours or power. Seen from a sociological perspective, they would, therefore, be no different from the office-bearers of any other small group, since their efficacy would be dependent on the goodwill of those in that group. This did not, however, give Rahner grounds to postulate that episcopal ministry might disappear. Indeed, he evinced no doubt that the believer of the future, far from seeing episcopal authority as oppressive, would accept that there must be 'a sacred order grounded in the Spirit of Christ'.[117] This was so because the bishop himself, as in the Church of the first centuries, would 'invite voluntary obedience and understanding for his decisions, in love and humility'.[118] In fact, Rahner suggested that the complex nature of life in the future might mean that many, or even most, concrete decisions were left to the conscience of the individuals.[119] Writing in 1965, he prophesied that the decision on birth-control might be included among such topics.[120] While not denying that the right to make binding decisions would still exist in the Church of the future, Rahner was also of the opinion that no one would be inclined to use it, especially in a paternalistic way which implied that all wisdom came from above—a point to be developed further in the next two chapters.

As the foregoing sections have shown, Rahner was deeply convinced of the benefits which Vatican II had brought to the Church—a fact which makes a remarkable contrast with his reflections on the state of the Church before the Council. At the same time, however, he was not unaware that the Council had been followed by much unhappiness and tension.

POST-CONCILIAR DIVISIONS

That Vatican II was the cause of turmoil in the Church came as no surprise to Rahner. He recognized that human beings would

[116] 'Episcopacy', *TI* vi. 368 (*ST* vi. 430–1).
[117] 'Future Reality', in *Christian of the Future*, 98 (*ST* vi. 495).
[118] Ibid. (ibid.). [119] Ibid. 98–9 (ibid.).
[120] 'Changing Church', ibid. 33 (ibid. 474).

always find it uncomfortable to live in an age of transition, where, as in the period after the Council, the old was passing away but the new had not yet become self-evident. None the less, Rahner urged that patience and courage, rather than succumbing to 'morbid nervousness', were the proper responses to such an upheaval.[121] As soon became evident, however, the possibilities for creativity in the post-conciliar Church were threatened by the excesses of both progressives and conservatives. Disclaiming extremism of any complexion, Rahner promoted the virtues of 'the centre'. Accordingly, he sought a path between the Scylla of Pian monolithism and the Charybdis of a 'Church' where everyone could believe and say what they chose.[122] Rahner's commitment to 'the centre' was not, however, universally shared.

Thus, as early as 1966, he found in a letter circulated to all episcopal conferences by Cardinal Ottaviani, then Prefect of the Congregation for the Doctrine of the Faith, a disturbing tendency towards imposing a homogeneous theology. Rahner feared that the success of such an endeavour would render the Church incapable of speaking to the contemporary pluralistic world.[123] A homogeneous theology could only be the theology of a sect. That his initial disquiet had not mistaken a passing phantom for a general trend was confirmed for Rahner in the years that followed. In 1979, for example, he expressed the opinion that the Congregation for the Doctrine of the Faith not only remained too fearful of new theology, but lacked creativity.[124] At the same time, he also voiced his concern that the Congregation's restrictive attitude would drive theologians to seek refuge in 'safe areas', like pastoral theology or religious education, rather than the more dangerous realm of systematics.[125]

Rahner's opposition to such centralizing trends did not, however, imply an uncritical endorsement of 'progressive' views. Indeed, in

[121] Ibid. 19 (ibid. 463).

[122] 'Teaching Authority', *TI* ix. 86 (*ST* viii. 115). The ref. to 'the centre' can be found on p. 100 (p. 132).

[123] Ibid. 90 (ibid. 119). The article begins with Rahner's summary of Cardinal Ottaviani's letter.

[124] 'Abiding Significance', *TI* xx. 95 (*ST* xiv. 309).

[125] Ibid. (ibid.). In this context, it is worth noting Rahner's opinion that Romano Guardini had concentrated on the philosophy of religion, rather than systematics, as a result of being 'traumatized' by what had occurred in the Church during the Modernist period, see *I Remember*, 74 (*Erinnerungen*, 81).

the very article in which he criticized the Congregation for the Doctrine of the Faith, he also expressed deep distrust of contemporary liberal proclivities. Of particular concern to him were those who wanted to remain in the Church even while they sought to reinterpret the Church's teaching in a way that substituted novelty for what was of permanent validity.[126] While such people would have previously left the Church, Rahner saw them now attempting to forge a place in it for their own approach to faith. This fashion he characterized as the unlikely marriage of an individualism which demanded the freedom to believe what one chose and a collectivism which none the less wanted a community which held the same opinions.[127]

As his response to such extremes, Rahner sought to make clear the parameters which defined his treasured 'centre'. On the one hand, he affirmed the right of the Church's teaching authority to say 'No' to certain propensities in theology, to indicate that they were not to be reconciled with Catholic faith.[128] Indeed, ever-sensitive to the impact of cryptogamic heresies, he cautioned both against a tendency towards secularism and desacralization within the Church itself, and against those heresies which, like the trend described above, weakened the substance of Christianity even while claiming to improve it.[129]

On the other hand, he was convinced that the teaching authority could not simply repeat inherited formulations of faith without attempting to reinterpret them to meet contemporary needs.[130] In addition, he stressed that the Congregation for the Doctrine of the Faith needed to realize that in a world where even the possibility of belief in God was consistently questioned, the issue of authority was not the only one the Church had to face—a theme which will feature prominently in the next chapter.[131] Accordingly, he urged those in authority to adopt a 'critical openness' which did not regard every dissenting thought as a threat.[132] While the Church wanted peace in its internal life, this could not be the

[126] 'Teaching Authority', *TI* ix. 90 (*ST* viii. 119).
[127] Ibid. (ibid. 120). [128] Ibid. (ibid.).
[129] Rahner with Johannes B. Metz, *Die Antwort der Theologen* (*Die Antwort*) (Düsseldorf, 1968), 14.
[130] 'Teaching Authority', *TI* ix. 90 (*ST* viii. 119–20).
[131] Ibid. 87–8 (ibid. 116–17). [132] *Die Antwort*, 17.

'peace of a graveyard' or of an 'indifferent conformism'.[133] Thus, neither an excess of control by the magisterium nor an intemperate disregard of the magisterium by theologians could guarantee the future well-being of the Church.

Even during the Council itself Rahner had argued that both progressives and conservatives needed to be aware not only that no single group had exclusive access to the truth, but that the future of the Church would ultimately be determined by the mysterious providence of God, not the programme of any particular 'party':

> The correct mixture and unity of the two attitudes (courage to risk decisions and to hear God's wisdom in the voice of the times; and a sobriety which does not identify one's own spirit simply with that of the Church) is ultimately always an imponderable which is not planned and clearly worked out by human beings, but appears within history as an accident of the constellation of history and which in faith must be accepted as divine providence.[134]

Finally, it is worth recording that Rahner regarded himself as a casualty of the divisions in the Church. In 1968, he acknowledged that though he had been regarded as a 'progressive' theologian in the twenty years before the Council, he now found himself defending the more traditional positions of the Church because it was those very doctrines which were being rejected by contemporary progressives. As a result, he felt that he had been metamorphosized from a 'leftist' to a 'rightist' not because he himself had actually changed his views, but because others had changed around him.[135] Yet, irrespective of how others might categorize him, Rahner continued to seek 'the centre'.

An aspect of this balance in Rahner's approach was his refusal to endorse the naïve hope that Vatican II would usher in the millennium. Indeed, he was convinced that, far from settling all issues, the Council had not only bequeathed many new tasks to the Church, but deliberately orientated the Church towards the unknown, and unknowable, future. Rahner's understanding of the nature of these tasks will be examined in what follows, the final section of this chapter.

[133] *Grace*, 35 (*Gnade*, 223).

[134] 'Die zweite Konzilsperiode', *Oberrheinishes Pastoralblatt*, 65 (1964), 81.

[135] *Die Antwort*, 13.

THE POST-CONCILIAR AGENDA

It was important for Rahner that Vatican II not be perceived as either having spoken the final word in theology or as making superfluous further change in the Church. Displaying a Churchillian touch, he stressed that the Council could not be regarded as anything more than 'the beginning of the beginning'.[136] There was, therefore, no place for a conciliar triumphalism. Indeed, even during the Council, Rahner had urged that Vatican II should not be seen as the culmination of the Church's mission: 'The Council is only a beginning. And this Council is only a service. Its ultimate aim is not the self-assertion of the Church in the future, but is the infinity of humanity and the arrival of the kingdom of God, which is, put simply: faith, hope and love.'[137]

Accordingly, he argued that the future ought to be dedicated not to the production of commentaries on the Council, but to carrying further what it had initiated. In this regard, the issue of the relationship of the episcopate and the primacy was not the only question which Vatican II had addressed but not fully resolved. There also remained a need for greater clarity concerning such themes as: the charismatic dimension of the Church; its missionary work; the place of the diaconate; and the 'hierarchy of truths'.[138] Rahner stressed that if the future desired to be faithful to the spirit, and not simply the letter, of Vatican II, its most important task would be to emulate the Council's method of confronting the times. To do so would mean abandoning the lagoons of theological subtleties in order to face radical questions such as whether belief was possible in a pluralist age—as will be seen in the next chapter, this topic was a major focus for Rahner in the 1960s and 1970s.[139] In addition to the basic questions associated with belief, another crucial issue which had been raised by contemporary history was the position of women within the Church.

Writing in 1964, Rahner urged patience with the Church as it struggled to come to terms with the radically altered role of women

[136] 'The Council: A New Beginning', in *The Church after the Council*, trans. D. C. Herron and R. Albrecht (New York, 1966), 20 (*Das Konzil: Ein neuer Beginn* (Freiburg i.B., 1966), 14).

[137] 'Die zweite Konzilsperiode', 82.

[138] 'Challenge to Theology', *TI* ix. 14–16 (*ST* viii. 27–9).

[139] Ibid. 18 (ibid. 29–30).

in society. Equally, however, he was insistent that the Church must address itself to the topic.[140] In his own discussion of the role of women, Rahner specifically referred to the declining numbers of ordained priests as an incentive to further reflection on the contribution women could make to the Church. Significantly, he did not thereby imply that women ought to be ordained to the priesthood—although he did acknowledge that the Church had an ancient, if controverted, tradition of deaconesses—but that they be enabled to exercise those ministries which properly belonged to them as laity; ministries which, therefore, did not make those who exercised them the mere subordinates or auxiliaries of the clergy.[141] Furthermore, he stressed that women in the Church ought not to wait for the hierarchy to decide their future but, precisely because they already enjoyed a share in the Spirit, ought to move towards devising their own concrete models of Christian living as women.[142]

As his approach to the position of women in the Church shows, Rahner was not to be found championing those who, because of either a romantic attitude towards the past or an aversion to the upheaval created by the new, promoted a more cautious approach to *aggiornamento*. Indeed, he stressed that the appropriate question for the post-conciliar period was not whether the Church was moving with reckless abandon, but whether its commitment to change was actually commensurate with that which existed in secular society, where the future had already begun.[143] In response to those who equated faithfulness with resistance to the new, Rahner reiterated his conviction, which was discussed at the beginning of this chapter, that a willingness to change could in fact be the deepest expression of the Church's faith:

> The Church is asked by God whether it has the courage to undertake an apostolic offensive into the future and therefore whether it has the necessary courage to show itself to the world in such an uninhibited way that no-one could have the impression that the Church goes on existing only as a relic from earlier times because it has not yet had enough time to die.[144]

[140] 'The Position of Woman in the New Situation in which the Church Finds Herself', *TI* viii. 77 ('Die Frau in der neuen Situation der Kirche', *ST* vii. 352–3).

[141] Ibid. 84 (ibid. 359).

[142] Ibid. 87–9 (ibid. 362–4).

[143] 'Changing Church', in *Christian of the Future*, 19 (*ST* vi. 463).

[144] Ibid. 36 (ibid. 477).

Rahner's perception of how its commitment to the world and to the future would reshape the Church will be the subject of the final chapter. What can be noted already, however, is his conviction that the future would not simply reproduce what had preceded it. One specific area where Rahner was convinced that the theology of the future would be different from what preceded it was ecumenism.

The primary challenge for the churches in the future, argued Rahner, would be to learn a new way of acting, a way of acting which focused on the future rather than on past divisions.[145] The nature of life in the diaspora called for changes in the style of both Catholic and Protestant theology: Catholics needed to avoid an exaggerated Mariology; Protestants needed to look beyond their own various competing theologies and beyond the ecumenical divide. While he encouraged a united effort to explain Christian faith to the people of tomorrow, Rahner none the less cautioned against the oversimplification of faith which he regarded as characteristic of the work of John A. T. Robinson—his specific criticism of the direction of contemporary theology will be detailed in the next chapter.[146]

Fundamental to Rahner's view of the future was the conviction that the balance between the changeable and the enduring could be found only by those who accepted that the whole truth was richer than their partial insights. To be avoided was the extremism which either believed that everything could be changed or refused to change anything.[147] Those who successfully negotiated such a challenge would realize the truth of the paradox that what endured had the strength to change, a strength derived from the plenitude of God.[148]

REVIEW

As has been indicated throughout this chapter, Rahner refused to accept that Vatican II marked the end of the Church's labours. Indeed, he believed that the real achievement of the Council was that it had committed the Church to a changing world. As interpreted by Rahner, this commitment implied the need for the Church to be aware that there were sources of knowledge outside

[145] 'Challenge to Theology', *TI* ix. 23 (*ST* viii. 36). [146] Ibid. (ibid.).
[147] 'Changing Church', in *Christian of the Future*, 37 (*ST* vi. 477).
[148] Ibid. 38 (ibid. 478).

specific Christian revelation, to be prepared to dialogue with those who had such knowledge, and to be willing to change in order that the Church's saving truth might be more effectively communicated to a changing world. For Rahner, this commitment to the world was not only more important than any single decree passed by the Fathers, it was also the key to the Church's future.

Vatican II's emphasis on the Church's relationship with the world thus shaped Rahner's belief that the main post-conciliar task lay not in gleaning from the Council's documents guide-lines for a new, 'permanent' model of the Church—which itself would have suggested that the Council, no less than the Pian era, had sought to immunize the Church against change—but in emulating the Council's willingness to move forward into the future which was not only unknown, but which could not be determined by any single force within the Church itself. Consequently, both Rahner's emphasis on the local Church and his description of the Church in the diaspora implied a Church characterized by variety rather than uniformity.

Rahner's firm conviction that the Church of the future would be a Church existing in the diaspora has attracted some criticism. Indeed, his view has even been categorically rejected as 'depressing speculation'.[149] While Rahner may have overstated the future smallness of the Church—as will be noted in the last chapter, Rahner's later view did not foresee as sharp a decline in numbers as his earlier articles had implied—Rahner's references to the diaspora were certainly not designed to induce depression. Most importantly, it must be remembered that Rahner did not predict the disappearance of Christianity, but rather the demise of 'the Catholic nation' and of 'folk-costume Christianity'. Since he was writing in a European context and against the particular contours of European history, where 'the Catholic nation' had been a reality, his ideas must first be situated in that context. Thus understood, it becomes more difficult to deny that his approach to the diaspora is an accurate descriptive of twentieth-century reality.

Furthermore, far from having merely negative overtones, the diaspora in Rahner's thinking was rich in potential for contributing to the Church's well-being. Thus, as was discussed in the

[149] Joseph H. Fichter, 'The Church: Looking to the Future', *America*, 160 (1989), 189.

previous chapter, he included it as a 'must' in terms of salvation history; a 'must' which was embraced by God's saving will. In addition, he argued that since the diaspora meant that membership of the Church could not be taken for granted, the Church of the future would be composed only of those whose genuine commitment to it would motivate them to contribute to its life. Ironically, it seems that on this point Rahner was actually too optimistic. While he was aware of the modern phenomenon of the 'marginal Catholic', Rahner may well have underestimated the extent to which such a relationship to the Church would remain a feature of modern life. While Rahner expected a firm decision for or against Church membership, he did not anticipate what is in fact characteristic of contemporary Church life: that many people retain contact with the Church only for occasions such as baptisms, weddings, and funerals.[150]

What has not been exhaustively examined in this chapter is how Rahner envisaged the Church's system of belief and the functioning of its structures would be affected by the commitment to the modern world which the Council initiated. As could be expected, however, from his desire to seek the 'centre', Rahner sought both to defend the value and truth of the Church's doctrine and structures and to adapt them to contemporary conditions. His proposals for achieving this balance will be the focus of the next chapter.

[150] See Neumann, 'Diasporakirche', 70.

6

Faith and Authority in a Pluralist Age

FUTURE historians writing about the twentieth century may find it difficult to understand the 1960s. How will they, for example, begin to make sense of 'psychedelia' or 'love-ins'? While these may prove puzzling, the great challenge will be to answer the questions raised by the values of the 'Sixties': Why was there such a profound dissatisfaction with the world which had been inherited? Why was there a widespread belief that the values of 'the older generation' had become moribund? Why was the great public activity of the Sixties—apart from the pop festival—the protest march, which combined 'anti-establishment' slogans with demands for a new world order? In short, these future researchers will have to account for the fact that, in everything from its music to its morality, the 'age of Aquarius' seemed to reject what had gone before.

Inevitably, their study will also have to explore the impact on religious faith of this upheaval. One way of doing this would be by examining the most controversial aspect of theology in the 1960s: 'secular Christianity'. This movement—a word which itself remains evocative of the loosely amalgamated groups whose campaigns embodied the reforming zeal of the Sixties—attempted to rethink theism, and Christianity in particular, in an environment radically different from that which had sustained 'classical Christianity'. In so doing, its protagonists—included among whom were John A. T. Robinson, often regarded as the symbolic figure of the movement, Harvey Cox, and Paul van Buren—developed an understanding of God, Christ, and the Gospel which was often at odds with Christian tradition.[1]

Although Karl Rahner's response to the challenge of the times

[1] For a sympathetic contemporary discussion of the origins and approach of 'secular Christianity', see Robert L. Richard, *Secularization Theology* (New York, 1967).

differed significantly from that of the mainly Protestant, and indeed Anglo-American, practitioners of the theology of secularism, he too was occupied in the 1960s, and beyond, with the relationship between Christian faith and modernity. Indeed, with no less ardour than Robinson himself, Rahner sought 'a radically new mould' of Christian belief and practice.[2] As has already been established in earlier chapters, however, Rahner was also insistent that the authenticity of the new depended on its continuity with the Church's history of faith.

While Rahner's quest for a Christianity attuned to the needs of the modern world increased in intensity during the 1960s, it did not begin in that decade of radicalism. As was seen in Chapter 4, his interest in the relationship between Christian faith and modernity had been evident even during the Second World War. That his concern with this theme was no fad is further confirmed by the central place it retained in his writings until his death in 1984. During this long period, however, his approach to the question did not remain static.

Whereas his reflections in the 1940s had concentrated on the impact of a changing social order on the Church's place in society and on the difficulty in maintaining orthodox belief in an age of cryptogamic heresies, the 'later' Rahner was concerned with the more fundamental question of whether belief in God was actually a realistic option in the second half of the twentieth century. As the source of the greatest challenge to the possibility of belief, Rahner nominated pluralism.

THE IMPACT OF PLURALISM

In Chapter 4, it was noted that Rahner used 'pluralism' to describe the peculiarly twentieth-century phenomenon of multiple and irreducible sources of knowledge. In that chapter, the emphasis was on the difficulty of making moral decisions in an environment characterized by such complexity. In the present chapter, however, the focus is broader. What will be examined here is how pluralism had affected the possibility of belief in God. As a prelude, it will be useful to reiterate what Rahner regarded as the primary trait of a pluralist world.

[2] The phrase comes from Robinson's *Honest to God* (London, 1963), 124. For details of Robinson's work, see Richard, *Secularization Theology, passim.*

As analysed by Rahner, the essence of pluralism was the existence of such a variety of possible knowledge that no individual could integrate it all into a unified world-picture.[3] Even without necessarily implying contradictions between various philosophies or sciences, the myriad sources of knowledge were such as to prevent practitioners of any one discipline being able to acquaint themselves with other disciplines to a degree sufficient to produce a world-view which integrated all available knowledge. One result of this trend was a proliferation of world-views. As a result of this proliferation, not only had the Christian world-view become one among many, but God was no longer universally invoked as proximate and ultimate explanation of everything. The movement away from a God-centred view of reality Rahner explored under the heading 'secularization'.

Put simply, secularization designated the process of interpreting the world in other than theistic terms. A radically non-theistic understanding of life owed its emergence to the developments within, and the multiplication of, the sciences. Not only had scientific progress demystified the world, but the empirical scientific method was not conducive to raising the metaphysical questions foundational to theism. Accordingly, claimed Rahner, it had become more difficult for people in the modern world to discover anything they could call God.[4] This difficulty was compounded by the fact that the level of certainty which believers had previously been able to achieve about the *praeambula fidei*—the 'pre-theological' supports for faith, among which was included acceptance of the possibility of 'natural revelation'—had also become a casualty of the advancing wave of secularization.[5] In addition to masking the 'footprints' of God in the natural world, secularization had even affected how society itself was construed. Thus, the invocation of God as both the ultimate explanation of

[3] 'A Small Question Regarding the Contemporary Pluralism in the Intellectual Situation of Catholics and the Church' ['Pluralism'], *TI* vi. 26 ('Kleine Frage zum heutigen Pluralismus in der geistigen Situation der Katholiken und der Kirche', *ST* vi. 40), orig. pub. 1965.

[4] 'Theological Considerations on Secularization and Atheism' ['Secularization'], *TI* xi. 169 ('Theologische Überlegungen zu Säkularisation und Atheismus', *ST* ix. 180), orig. pub. 1968.

[5] 'On the Situation of Faith' ['Situation'], *TI* xx. 24 ('Zur Situation des Glaubens', *ST* xiv. 36), orig. pub. 1980. For a fuller explanation of the *praeambula fidei*, see Rahner and H. Vorgrimler, *Dictionary of Theology*, 2nd edn., trans. R. Strachan, D. Smith, R. Nowell, and S. O'Brien Twohig (New York, 1985), 402–4 (*Kleines theologisches Wörterbuch*, 10th edn. (Freiburg i.B., 1976), 343–4).

all reality and the sole basis of stability in society had been replaced by reference to what human beings could themselves achieve via a liberal education, social pressures, and a necessary interdependence.[6]

In depriving belief in God of much which had traditionally underpinned it, secularization had inevitably weakened the foundations of the Church's position in society: a world beginning to define itself without reference to God clearly had little need for a Church claiming to be the authoritative interpreter of the word of that same God. Consequently, not only was the Church no longer regarded as alone holding the keys to human progress, but there was an increasing separation between the Church and the world.[7] The impact of pluralism on theism and on the Church sets the agenda for the following sections: in addition to examining Rahner's defence of belief in God in the modern world, what will also be explored is the argument he advanced to support the value of an ecclesial dimension for such belief.

THE POSSIBILITY OF BELIEF

A comprehensive description of Rahner's defence of theism would require expounding the development of his fundamental theology in the years after the Council, but this will not be attempted here. Instead, what follows has two more modest aims: to identify the formal principle which, as early as 1965, Rahner used to justify belief; to outline Rahner's notion that the impact of secularization, far from being thoroughly negative, could actually assist in clarifying the nature of humanity's relationship to God.

Although he acknowledged the seemingly limitless complexity of the modern world, Rahner rejected the notion that belief became possible only when every pertinent question or objection had been answered definitively. In place of such a process, Rahner suggested something akin to Newman's use of the 'illative sense': an indirect proof arrived at through induction.[8] Without denying

[6] 'Secularization', *TI* xi. 173 (*ST* ix. 184–5).

[7] 'Theological Reflections on the Problem of Secularization' ['Problem of Secularization'], *TI* x. 319 ('Theologische Reflexionen zur Säkularisation', *ST* viii. 637–8), orig. pub. 1967.

[8] For Newman's own description of the dynamics of this 'illative sense', which allowed a judgement based on converging probabilities, see his major study, first pub. 1870, *An Essay in Aid of a Grammar of Assent* (Notre Dame, Ind., 1979),

that theists had the ongoing task of attempting to meet those objections to belief raised by various sciences and philosophies, Rahner argued that, by relying on the convergence of probabilities, the individual could nevertheless achieve a level of certainty morally and ethically sufficient to justify belief.[9] Such an approach he regarded as characteristic of the history of belief, as even uneducated people in the past had seen themselves as having adequate grounds for belief.

In thus asserting that faith was possible even in a secularized world, Rahner was not merely entertaining a pious hope, but applying the principles of his anthropology. Indeed, as was discussed in Chapter 1, the heart of Rahner's theology was his belief that humanity's transcendental reference to the mystery of God was the most fundamental fact about humanity. Since this relationship with God was constitutive of humanity, it followed that no matter what the social situation in which they lived, human beings could never be fully severed from their link to God.

Furthermore, Rahner argued that since the Incarnation had revealed not the dominance of the natural world over humanity, but that humanity was God's deputy in regard to nature, a less numinized view of the world did not necessarily indicate that human beings were usurping God's role in the world, but that they were actually exercising their God-given potential to understand and develop the world.[10] Consequently, it was possible to regard contemporary developments positively:

> It is inevitably true, and in an ever-increasing measure that the person of today and tomorrow is the person who has really become a subject, who is responsible for her- or himself not merely in a theoretical-contemplative sense, but actually; who has achieved the Copernican shift from 'cosmocentrism' to 'anthropocentrism' not merely in a theoretical and religious sense but actually.[11]

270–81. See also Bacik, *Apologetics and the Eclipse of Mystery*, 108–9; and Heinrich Fries, 'Theologische Methode bei John Henry Newman und Karl Rahner', *Catholica*, 33 (1979), 109–33, esp. 124–5.

[9] 'Pluralism', *TI* vi. 30 (*ST* vi. 44). See also *Foundations of Christian Faith: An Introduction to the Idea of Christianity* [*Foundations*], trans. W. V. Dych (New York, 1978), 1–14 (*Grundkurs des Glaubens: Einführung in den Begriff des Christentums* [*Grundkurs*] (Freiburg i.B., 1976), 13–25).

[10] 'The Man of Today and Religion', *TI* vi. 10 ('Der Mensch von heute und die Religion', *ST* vi. 21–2). Although a version of this article appeared in 1962, it was not pub. in its present form until 1965.

[11] Ibid. 8–9 (ibid. 19–20).

Even though he accepted that atheism was certainly a possibility if human ingenuity was misrepresented as conferring independence from God, Rahner did not subscribe to the view that secularization necessarily involved an 'a-theistic profanation' of the world.[12] Indeed, consistent with his affirmation of Vatican II's refusal to assert that the Church ought to control the secular world, he encouraged the Church to allow 'the worldly world to be really worldly'.[13]

While secularization had not obliterated—indeed could not obliterate—humanity's relationship to God, Rahner accepted that it probably meant the Christian message would remain ineffective for the majority of people in any future world. Consequently, it was not an articulated theism, but an 'anonymous' response to God's offer of salvation which was likely to become the norm.[14] If this was so, then the survival of a faith lived out in the Church seemed mortally imperilled. Whether Rahner himself subscribed to such a view will be discussed in what follows.

THE VALUE OF ECCLESIAL FAITH

Far from predicting that the Church would wither away, Rahner suggested that the prospects of a Christianity embodied in the Church might actually be more healthy than those of an abstract theism.[15] Such a claim clearly supposed that the Church, even a Church affected by the sinfulness of its members, had, precisely *qua* institution, something indispensable to offer. For Rahner, this 'something' was freedom from subjectivism, a freedom which was a crucial support for truth. This freedom was particularly to be prized in the context of the ultimate questions with which religion was concerned:

> If religion involves what is most real and complete in a person, then it cannot a priori be merely what is individual and the innermost reality of the individual person alone. Religion must be my own proper and free conviction, must be capable of being experienced at the very heart of

[12] 'Problem of Secularization', *TI* x. 319 (*ST* viii. 638). See also Patrick J. Lynch, 'Secularization Affirms the Sacred: Karl Rahner', *Thought*, 61 (1986), 381–93.

[13] *Handbuch der Pastoraltheologie*, ii/2. 35.

[14] 'Secularization', *TI* xi. 176–7 (*ST* ix. 187–8).

[15] Ibid. 177 (ibid. 188).

existence. But this existence is itself found only in a community and society, being revealed in giving and receiving.[16]

For Rahner, truth, especially when it related to the fundamentals of human existence, was inextricably linked with institution. Human beings, he argued, arrived at truth via a common search and dialogue: only through encountering the opinion of others could we be sure that our truth was more than self-deception. Indeed, only through such encounter did our truth really become our truth. The alternative was the 'hell of absolute aloneness'.[17] The Church, therefore, like other institutions, represented the objectivity of reality; through it, 'the other' had a meaning for me and was beyond my manipulation. The institution, therefore, showed that truth was actually, and not simply ideally, independent of me.[18] While this was true of institutions in general, it had a particular application to Christianity: 'Christianity is the religion of a demanding God who summons my subjectivity out of itself only if it confronts me in a church which is authoritative'.[19]

To this principle of the necessary relationship between truth and institution, Rahner added a further principle that dealt specifically with Christian revelation. Thus, he argued that faith in Christ as 'the absolute bringer of salvation' implied faith in the continuation of a community believing in him.[20] Furthermore, belief in the indestructibility of the Church's faith suggested that propositions articulating the heart of that faith also shared in its indestructibility. Without such propositions, there could be no historical connection between Christ and the faith of the community.[21]

While Rahner can thus be said to have regarded membership of the Church as an unequivocal support for faith, it is crucial to note that he was not thereby claiming either that the Church was the source of an individual's faith or even that faith in the Church was easily achieved. Although he did in fact prize the link between truth and institution, Rahner was far from suggesting that institutions could either substitute for personal convictions or could

[16] 'Courage for an Ecclesial Christianity', *TI* xx. 9 ('Vom Mut zum kirchlichen Glauben', *ST* xiv. 18), orig. pub. 1979.

[17] *Grace*, 171 (*Gnade*, 133). [18] Ibid. 171–2 (ibid. 133–4).

[19] *Foundations*, 344 (*Grundkurs*, 334).

[20] 'Does the Church Offer any Ultimate Certainties?' ['Certainties'], *TI* xiv. 55–6 ('Bietet die Kirche letzte Gewißheiten?', *ST* x. 294–5), orig. pub. 1972.

[21] Ibid. 57 (ibid. 296).

demand uncritical obedience. In addition, he accepted that in a world which had long since rejected absolutism, the authority of institutions, the Church included, had itself become problematic. Indeed, as Rahner wryly observed, it was easier to believe in the Pope when there were also kings whose role in society was beyond dispute.[22] In both contemporary Church and State, however, those who exercised authority were subject to critical scrutiny from their constituents:

People today have a different, prosaic and more or less demythologised attitude to authorities in general and to those who possess authority, since their history with its upheavals and mistakes is more obvious to them than it was to people in earlier times, and hence they know also about the mistakes and wrong decisions in the history of the Church's magisterium.[23]

Ironically, in thus making faith in the Church less than a self-evident necessity, the pluralist age highlighted the proper relationship which existed between faith in God and faith in the Church. As was discussed in Chapter 1, Rahner stressed that faith in the Church was both derived from, and therefore secondary to, faith in God. This distinction had been blurred in the period when membership of the Church was an unquestioned consequence of social existence. Although he did not doubt that it was none the less possible for an individual to have faith in the Church, Rahner stressed that such faith was not the primary feature of a person's life. More fundamental were a basic trust in the meaning of existence and a hope for salvation. These were both logically and psychologically prior to acceptance of the authority of the Church. No authority in the Church could substitute for this primordial faith:

A person will naturally find this ultimate, fundamental trust easier in a community of believers and so we can say that it has, even when life is understood concretely, an ecclesiastical element. But this does not alter the fact that, logically and psychologically, this fundamental trust comes before belief in the Church's authority, and supports, rather than being supported by, belief in the teaching authority of the Church.[24]

Rahner did not, therefore, portray faith in the Church as the indisputable foundation of belief in either God or the meaningfulness

[22] 'Situation', *TI* xx. 23 (*ST* xiv. 35). [23] Ibid. 19 (ibid. 31).
[24] 'Certainties', *TI* xiv. 49–50 (*ST* x. 288–9).

of existence. Indeed, he argued that in contemporary society ecclesial faith was not without its risks. This was so because the Church was often experienced as a burden rather than as the source of liberation.[25] The burdensome nature of ecclesial faith derived from the fact that Church officials, despite appearing 'pitiful, narrow-minded, old-fashioned, and out of date' could legitimately make demands on the believer in the name of the Gospel.[26] The possible aversion to the Church resulting from the inadequacies of its leaders was compounded by the fact that the modern believer was also aware of, and confronted by, the Church's 'often terrifying history'.[27]

Given his constant reference to the Church's sinfulness, it might well appear that Rahner regarded the Church more as a hindrance than a support to faith. As a general principle of human behaviour, however, he argued that we ought reject what has been meaningful only when it could be replaced by something more meaningful.[28] Since those who abandoned the Church would have only an abstract God and an abstract Christ as projections of their subjectivity, even a burdensome ecclesial faith remained infinitely preferable. This principle notwithstanding, there remained the question of how believers ought to respond to the unattractive and repellent in the Church's past and present.

As he had based himself on Newman in accepting the 'convergence of probabilities' as sufficient grounds for belief in God, Rahner again drew on that same Englishman to promote a positive response to the Church. While accepting that all the difficulties concerning the Church would not be resolved, Rahner followed Newman in claiming that even a thousand such difficulties did not necessarily amount to an insuperable doubt.[29] What was required, therefore, was not rejection of the Church, but entrusting oneself to it in an act of love. As with any human love, such love only became real when it involved the willingness to surrender

[25] 'Situation', *TI* xx. 21 (*ST* xiv. 33).

[26] 'Concerning our Assent to the Church as She Exists in the Concrete' ['Assent'], *TI* xii. 145 ('Über das Ja zur konkreten Kirche', *ST* ix. 482), orig. pub. 1969.

[27] 'Situation', *TI* xx. 21 (*ST* xiv. 33).

[28] 'Assent', *TI* xii. 151 (*ST* ix. 488).

[29] Ibid. 150 (ibid. 487). For Newman's ref. to 'ten thousand difficulties' not making a doubt, see *Apologia Pro Vita Sua* (London, 1976) 160. The orig. edn. of the *Apologia* was pub. 1864.

oneself without having fully understood the beloved.[30] To be avoided, was scepticism. Just as the possibility of disappointment in human relationships did not justify a permanent withdrawal from one's fellow human beings, so too what the Church had to offer in terms of its witness to faith was not to be spurned because the Church did not radiate in an unambiguous way the God who animated it. As the Cross revealed its richness only to those not distanced by its scandalous form, so too the Church could be experienced as the obvious centre of faith, where God's pledge of life in Christ was made tangible, only by those willing to enter it.[31]

This general affirmation of the Church's value leaves unspecified, however, the contours of the relationship between the believer and the Church in a pluralist age. In exploring this relationship, Rahner was particularly concerned to elaborate how the faith of the individual would relate to the corpus of belief articulated and guaranteed by the Church's teaching authority. His approach to this issue will be the next topic.

THE FAITH OF THE INDIVIDUAL AND THE FAITH OF THE CHURCH

To those who identified assent to a fixed body of teaching as constitutive of the Catholic faith, the pluralist age dealt a crushing blow. Indeed, Rahner stressed that not only the absolute assent of faith, but also the possibility of an authoritative, indisputable statement embodying that faith, seemed alien in a world dominated by a sceptical relativism which regarded even secular knowledge as merely provisional.[32] In a homogeneous Christian society, the Church's doctrinal structure was a source of strength; in a secularized world, it became one of the obstacles to faith. In addition, the irreducibility of world-views brought about by pluralism impinged on Catholics as much as on any other members of society. Accordingly, what one believed as a Catholic could often not be adequately synthesized with secular truths.[33] Moreover, the fact that Catholics existed as a minority group in society

[30] 'Assent', *TI* xii. 153 (*ST* ix. 490).
[31] 'Situation', *TI* xx. 21 (*ST* xiv. 33–4). [32] Ibid. 23 (ibid. 35).
[33] 'The Faith of the Christian and the Doctrine of the Church' ['Doctrine'], *TI* xiv. 30 ('Der Glaube des Christens und die Lehre der Kirche', *ST* x. 268), orig. pub. 1972.

was a further consequence of pluralism. Indeed, in an article written in 1975, Rahner coined the phrase 'gnoseological minority' to describe the status of Catholics in a world where their views were generally not shared by their neighbours.[34]

Since the believer had to struggle with conflicting world-views, Rahner acknowledged that the life of the Catholic in such a situation could be both painful and perilous.[35] Such difficulties did not, however, justify the adoption of any facile 'solutions' directed towards a more comfortable existence. Thus, he cautioned against any willingness either to deny or simply to accept the intractability of pluralism—an intractability which gave birth to a 'gnoseological concupiscence', since it could neither be eliminated nor synthesized.[36] Instead, what was to be aimed at was an 'asymptotic integration' of Christian faith and the knowledge generated by the plethora of contemporary philosophies and sciences.[37] This project required, however, an eschatological orientation. Not surprisingly, the prospect of an eschatological resolution of gnoseological concupiscence did not satisfy the immediate desire to know that one's faith was built on a sure foundation. Nevertheless, there seemed little likelihood that the believer could attain such certainty in a pluralist age, especially since the authority of the Church's magisterium, the traditional guarantor of such certainty, had itself become problematic.

In response to these developments, Rahner not only sought to provide reassurance that firmness of faith was not a chimera, he also continued to insist on a legitimate, indeed indispensable, role for the teaching authority. As will be seen, however, he did not imply thereby either that the level of certainty attained by the individual would fully reflect the articulated faith of the Church or that the future activity of the magisterium would merely reproduce what had been done in the past.

In fact, Rahner accepted it as inevitable that the difficulty of attaining faith within a pluralistic milieu would result in a gap between the level of faith achieved by an individual and the fully articulated faith of the Church:

[34] 'Transformations in the Church and Secular Society' ['Transformations'], *TI* xvii. 168–9 ('Kirchliche Wandlungen und Profangesellschaft', *ST* xii. 514).

[35] 'Doctrine', *TI* xiv. 36 (*ST* x. 275). [36] Ibid. 34 (ibid. 273).

[37] 'Philosophy and Philosophising in Theology' ['Philosophy'], *TI* ix. 52 ('Philosophie und Philosophieren in der Theologie', *ST* viii. 73), orig. pub. 1968.

Precisely today, when it is a person's most ultimate decisions with regard to God, Jesus Christ, and the hope of eternal life that are in question, it is quite impossible, and moreover not for one moment to be supposed in the case of an average Christian, even one who is interested in religion and well instructed, that his or her personal commitment of faith will be equally alive and vigorous to every truth without exception.[38]

Furthermore, he argued that in such a context it was actually justifiable for all believers, in order to preserve a basic certainty of faith, to ignore those aspects of magisterial teaching with which they felt unable to cope in their concrete situation.[39] The only proviso was that this method not be used in order to resist being led by God into new depths of truth. Consequently, each individual believer was free to develop a personal hierarchy of truths in order to maintain a firm conviction of God's offer of salvation in Christ. The multiplicity, indeed pluralism, of expressions of faith which would thereby result occasioned Rahner no anxiety. In fact, he accepted as proper that there should be a distinction between an objective hierarchy of truths and a personal one, determined by the individual's concrete conditions.[40]

In the decade following the Council, Rahner himself composed a series of 'short formulas' which were designed to show that it was possible to devise authentic expressions of Christian faith which were also attuned to the pressures exerted within a pluralist society. His efforts at such formulations had two aims: by separating the core of Christian faith from its secondary aspects, to make it possible for potential believers to distinguish between the fundamentals of Christianity and the 'uninviting and repellent' in the Church's appearance[41]; and to provide a formulation which would enable contemporary Christians to take responsibility for their faith in a non-Christian world, a responsibility which could be shouldered only when believers had a firm grasp of faith.[42]

[38] 'Heresies in the Church Today?' ['Heresies Today'], *TI* xii. 126 ('Häresien in der Kirche heute?', *ST* ix. 462), orig. pub. 1968.

[39] 'Doctrine', *TI* xiv. 38–9 (*ST* x. 276–7).

[40] 'Reflections on the Problems Involved in Devising a Short Formula of the Faith' ['Devising a Short Formula'], *TI* xi. 236 ('Reflexionen zur Problematik einer Kurzformel des Glaubens', *ST* ix. 248), first pub. 1970.

[41] 'The Need for a "Short Formula" of Christian Faith' ['Short Formula'], *TI* ix. 117 ('Die Forderung nach einer "Kurzformel" des christlichen Glaubens', *ST* viii. 153). The orig. form of the article was pub. in 1965; in its present form, it first appeared in 1967.

[42] 'Devising a Short Formula', *TI* xi. 230–1 (*ST* ix. 242).

To be successful, such short formulas needed to resemble advertising slogans or the policy statements of political parties: to be brief and self-explanatory.[43] They were to be not merely short, but also short-lived, to be reformulated as society itself developed. They would not, therefore, have the longevity of the Apostles' Creed. While Rahner was convinced that the Apostles' Creed would remain a permanent and binding norm of faith, he was equally certain that it could not function as a brief formula, since it was alien to the contemporary intellectual situation in which even 'God' could not be assumed to be comprehensible.[44] Similarly, although applauding the aim of Paul VI in producing a modern *Credo*, Rahner found the document itself flawed, both because of its excessive detail and because of its reliance on classical scholasticism.[45]

Far from being merely intellectual elucidations of the notion of 'God', the short formulas were to express the relationship between God and human existence. Rahner's emphasis on the existential dimension stemmed from his conviction that if theology was so divorced from modern life that it made the content of revelation unworthy of belief, then the very fact of revelation would also seem incomprehensible.[46] Despite the failure of much theology to address modern concerns, Rahner remained convinced that it was possible both to be existentially relevant and to capture the central truths of Christian faith:

> A brief formula must on the one hand express the basic substance of the reality of the Christian faith in such a way that the most intelligible possible access to it is opened up at the level of people's experience in the concrete conditions of their existence; while on the other hand, this basic substance can certainly be discovered in the idea of God turning to the world in the Trinitarian dimension of the economy of salvation.[47]

[43] 'Short Formula', *TI* ix. 120–1 (*ST* viii. 157–8).

[44] 'Devising a Short Formula', *TI* xi. 231 (*ST* ix. 243).

[45] Ibid. (ibid. 242). Paul VI's 'Credo of the People of God' can be found in Austin Flannery (ed.), *Vatican Council II*, ii (New York, 1982), 387–96. The orig. is in *AAS* 60 (1968), 433–45. The document was first pub. on 30 June 1968.

[46] 'Theology and Anthropology', *TI* ix. 41 ('Theologie und Anthropologie', *ST* viii. 60), orig. pub. 1966.

[47] 'Devising a Short Formula', *TI* xi. 244 (*ST* ix. 256). Rahner's own early attempt to devise formulas on a Trinitarian basis can be found in 'Short Formula', *TI* ix. 121–6 (*ST* viii. 158–64); in 'Devising a Short Formula', *TI* xi. 237–44 (*ST* ix. 248–56), he developed further the theological priorities which such formulas ought to respect. Rahner's final effort, written in 1976, at composing short formulas can be found in *Foundations*, 448–59 (*Grundkurs*, 430–40).

That even the short formulas did not completely bridge the gap between the individual's hierarchy of truths and the complete corpus of the Church's doctrine, Rahner regarded as cause for neither scandal nor panic. Indeed, his analysis of the act of faith, expounded in both his 1972 article, 'The Faith of the Christian and the Doctrine of the Church', and in 'The Situation of Faith', written in 1980, emphasized that to be excessively concerned about such a gap was to misunderstand the deepest meaning of faith. The basic point insisted on by Rahner was that faith was a single movement of the human person in God—grace—to God—eternal life.[48] Faith, therefore, designated the human response to God's self-communication. Consequently, the various 'objects of faith', the Church's expressed dogmas, had meaning only when related to the primordial response to God.[49]

Rahner asserted that the human act of believing was neither exhausted nor indeed fully understood if described only in terms of choosing to accept a particular categorial object. Faith was not, therefore, just one decision after another—'I will believe this, but I will not believe that'—but rather expressed the capacity of human subjects, via a basic decision, to dispose of themselves as a whole.[50] He stressed that since such an absolute decision was itself a response to the supernatural existential—that is, to that offer of God which was constitutive of human existence—it could properly be regarded as theological faith. In concrete terms, 'faith' was thus the exercise of a fundamental option for or against God, rather than assent to specific aspects of Church teaching. In addition, since the need to make such a basic decision was an integral aspect of being human, faith became a possibility for all, including those without a specifically religious world-view.[51]

If it was the fundamental option for God which was constitutive of faith, then assent to specific doctrines was indeed secondary. While this approach may have resolved the dilemma of how faith could be present without specific acceptance of the Church's teachings, it none the less exposed Rahner to the charge of failing to respect the traditional distinction between the *fides qua*, the fact that one believed, and the *fides quae*, the content of such faith. Indeed, his approach seemed to imply that someone could claim

[48] 'Doctrine', *TI* xiv. 39 (*ST* x. 278). [49] Ibid. 39–40 (ibid. 278–9).
[50] 'Situation', *TI* xx. 27–8 (*ST* xiv. 41–2). [51] Ibid. 28 (ibid. 42).

to be a believing Catholic even without giving absolute assent to any of the Church's doctrines.

In defence of his position, Rahner asserted that *fides qua* and *fides quae* were in fact identical in their origins. This was so because the fundamental reality believed in—God—was also the principle which sustained faith.[52] Thus, the absolute assent constitutive of faith was not directed towards the subject's 'objectless transcendence', but towards God.[53] Consequently, it could be claimed that both the individual and the Church lived by the same basic act of faith.[54]

In addition, the encounter with God's immediacy, which was the occasion calling forth the subject's absolute assent, did not take place at the level of mystical interiority, but through a posteriori realities: other persons or the environment. These objective realities Rahner regarded as categorial objects of revelation—although 'revelation' did not imply that they were either revealed for their own sake or even regarded as having a particular religious meaning, but that they were the means by which the individual encountered God.[55] In the light of such an interpretation of revelation, it was possible to conclude that the radical human decision involved in accepting God's self-communication was not simply *fides qua*, but also *fides quae*. In addition, Rahner was convinced that only such an interpretation could justify Vatican II's optimism about the possibility of salvation for pagans and atheists.[56] This interpretation also opened the way for Rahner to propose a new definition of the *depositum fidei*, a definition which—consistent with the ideas discussed in Chapter 3—identified the content of faith with the Spirit:

The *depositum fidei* is not primarily and originally a number of propositions composed by human beings, but is the Spirit of God, which had been irrevocably communicated to humanity and brings about in the concrete person the saving faith which that person actually has. Of course, this Spirit also brings about the community of believers in which the

[52] 'Doctrine', *TI* xiv. 39 (*ST* x. 278).
[53] 'Situation', *TI* xx. 29 (*ST* xiv. 43).
[54] 'Doctrine', *TI* xiv. 40 (*ST* x. 279).
[55] 'Situation', *TI* xiv. 30 (*ST* xiv. 44–5). See also his 'Observations on the Concept of Revelation', in Rahner and Joseph Ratzinger, *Revelation and Tradition*, trans. W. J. O'Hara (Freiburg i.B., 1966), 9–25 ('Bemerkungen zum Begriff der Offenbarung', in *Offenbarung und Überlieferung* (Freiburg i.B., 1965), 11–24).
[56] 'Situation', *TI* xiv. 30 (*ST* xiv. 45).

unity of their faith and its complete fullness objectifies itself and manifests itself reflexively, which we call the official belief of the institutional Church. This does not, however, alter the fact that, from beginning to end, it depends on the faith actually present in the concrete Christian, a faith which is salvific in its fullest sense and communicated from God, no matter how slight and fragmentary might be its objectification in the mind of a person.[57]

Thus, as he had defined a 'believer' as one who accepted God's self-communication, even though such faith might never be made explicit, so Rahner accepted that being a Catholic Christian did not imply the necessity of absolute assent to all the Church's dogmas. Indeed, he refrained from fixing even a necessary minimum of beliefs to be expressly held by members of the Church. Instead, he argued that what was essential was the recognition that 'absolute mystery', a mystery which had spoken to us in Christ, was at the centre of one's life.[58] Beyond that, what was needed for a baptized person to continue to be regarded as a Catholic was that they did not trespass against the traditional bounds defining heresy, that is, they did not 'explicitly and publicly' deny a particular dogma.[59] While individual members would thus fail to achieve a level of faith equal to that of the Church as a whole, Rahner stressed that all nevertheless shared in the further development of faith which was present at the collective level.[60]

In discussing the gap between the faith of the individual and the faith of the Church, Rahner specifically addressed himself to the situation of those who, despite being baptized, either had little contact with the Church or had doubts about the substance of Catholic faith. He argued that such 'partial identification', when it derived from difficulties with belief, was not to be immediately equated with subjective guilt. This was so because a denial of certain elements of the Church's faith could coexist with a *fides implicita* which actually affirmed the spirit of the Church's law and teaching.[61]

[57] 'What the Church Officially Teaches and What the People Actually Believe', *TI* xxii. 169–70 ('Offizielle Glaubenslehre der Kirche und faktische Gläubigkeit des Volkes', *ST* xvi. 223), orig. pub. 1981.

[58] 'Doctrine', *TI* xiv. 46 (*ST* x. 285).

[59] Ibid. 24 (ibid. 262).

[60] Ibid. 40 (ibid. 279).

[61] 'Schism in the Catholic Church?' ['Schism'], *TI* xii. 113–14 ('Schisma in der katholischen Kirche?', *ST* ix. 449), orig. pub. 1969.

Furthermore, Rahner stressed that it was impossible to know whether such 'borderline Catholics' only had half their faith left or had only acquired half their faith. If it was the latter which was true, if such people were 'centripetal' rather than 'centrifugal' Catholics, then their partial identification was appropriate to their stage of development.[62] In addition, he argued that the existence of differences in both world-view and faith consciousness even among those who were 'practising Catholics' provided yet another reason not to be precipitate in condemning those on the edge of the Church.[63] While Rahner's emphasis on the primacy of personal faith was clearly sympathetic to those who had difficulty with the Church's teaching, what needs to be investigated further is whether it actually left a role for a structured teaching authority in the Church.

THE ROLE OF THE MAGISTERIUM

In discussing Rahner's understanding of the teaching authority in the Church, it is necessary to distinguish between a general description of its functions, one applicable to all periods of history, and Rahner's specific reflections on how the pluralism of the twentieth century had affected both the style and substance of the ministry exercised by the magisterium.[64] To begin, the emphasis will be on what Rahner accepted as a permanent duty of the teaching authority: presiding over the Church's unity of faith.

As an abiding principle, Rahner accepted, as was noted above, the need for infallible propositions defining the community's faith. While the Church's unity could not be reduced simply to a matter of words, it was none the less true that statements expressing its shared faith were indispensable. Their contribution to the maintenance of this unity could not be replaced by any common sympathy or even any common action or cult.[65] The responsibility for

[62] 'On the Structure of the People of the Church Today' ['People of Church'], *TI* xii. 224–5 ('Zur Struktur des Kirchenvolkes heute', *ST* ix. 564–5), orig. pub. 1970.

[63] Ibid. 226 (ibid. 566).

[64] Throughout the following sections, 'teaching authority' and 'magisterium' will be used interchangeably when referring to those in the Church empowered to define doctrine. Rahner himself relied exclusively on the term *das Lehramt*.

[65] 'Pluralism in Theology and the Unity of the Creed in the Church' ['Creed'], *TI* xi. 21 ('Der Pluralismus in der Theologie und die Einheit des Bekentnisses in der Kirche', *ST* ix. 30), orig. pub. 1969.

formulating such propositions belonged, exclusively, to the teaching authority—the fence of nuances with which Rahner surrounded this right will be added shortly. In addition, Rahner stressed that the Church's unity of faith also required that such propositions be understood in the way specified by the teaching authority—this is the idea of a *Sprachregelung* which will be a major element of this chapter.[66]

In regard to the binding nature of the Church's teaching, Rahner reiterated that, in order to preserve the unity of the community of faith, Christian tradition—which had been reaffirmed in *Lumen Gentium* (*LG* 12)—regarded as incompatible with Church membership any unequivocal rejection of those statements which the Church, represented through the teaching authority, proclaimed as an element of God's revelation in Christ.[67] When propounded with this level of solemnity, magisterial statements acted as a border delineating what could be claimed as a legitimate expression of Catholic faith—the status of those magisterial statements lacking such a total commitment of the teaching office will also be a future topic. Those who either consciously and deliberately rejected such statements or proposed alternatives which the teaching authority judged irreconcilable with them thereby ceased to belong to the Church. Rahner argued that even prior to being officially anathematized as heretics, such people had in fact separated themselves from the Church, since they could no longer be considered members of 'a spiritual community sharing the same faith'.[68]

Not only the Church's existence as a community of faith, but also the fact that it was a social entity required that the Church protect itself against abuses threatening its foundations.[69] In fact, Rahner stressed that the Church would be betraying both the Gospel and itself were it to lack the courage to pronounce a clear 'No' to opinions that could not be united with its faith.[70] Consistent with the view he had expressed long before Vatican II, Rahner argued that although the Church was to be a bulwark of freedom, this did not commit it to granting the freedom to undermine the

[66] 'Heresies Today', *TI* xii. 132 (*ST* ix. 468).
[67] 'Doctrine', *TI* xiv. 25 (*ST* x. 263).
[68] 'Heresies Today', *TI* xii. 125 (*ST* ix. 461).
[69] 'Doctrine', *TI* xiv. 28 (*ST* x. 266).
[70] 'Heresies Today', *TI* xii. 136 (*ST* ix. 473).

Church: 'The sphere of freedom of conscience and opinion, which the Church itself protects and recognises, is not simply coterminous with the sphere of that which can be believed and taught in the Church.'[71]

Rahner stressed, however, that the exclusion of heretics related only to the Church's socio-juridical nature. Thus, when the Church rejected a particular opinion as unacceptable, it was not thereby making a statement about how those who propounded it stood in relationship to the mystery of grace which the Church represented.[72] Moreover, Rahner's defence of the Church's right to expel heretics was not a refutation of his defence of the legitimacy of the gap between the faith of the individual and that of the Church. Indeed, he reaffirmed his conviction that, for many contemporary Catholics, aspects of the Church's doctrine, such as the emphasis on Mary or even the sacraments, might remain alien.[73] Not sharing the fullness of the Church's faith was blameworthy only when it became a denial of what was not fully understood.

Returning to his emphasis on the link between truth and institution, Rahner argued that the magisterium's authoritative definition of the faith of the Church functioned as an antidote to the subjectivism which was often attendant on pluralism. He stressed that truth often seemed alien because it challenged the capriciousness of our subjectivity.[74] An external authority could, therefore, act as an 'archimedean point' which was beyond our manipulation.[75] Indeed, Rahner suggested that those who definitively rejected the Church's teaching were often the very ones who failed to be aware of the danger of subjectivity:

> In the eyes of those who are self-critical and modest, must that which they fail to understand always straightaway be rejected as that which is false? If they did absolutely identify themselves with their own subjective feelings at any particular moment, would not such an attitude be the death of all delight in intellectual growth?[76]

As has been alluded to in a number of places throughout this section, Rahner's advocacy of the need for a teaching authority to preserve the unity of the Church's faith was not without significant

[71] Ibid. 133 (ibid. 469–70). [72] 'Doctrine', *TI* xiv. 28 (*ST* x. 266).
[73] 'Heresies Today', *TI* xii. 127 (*ST* ix. 463).
[74] 'Assent', *TI* xii. 157 (*ST* ix. 494). [75] Ibid. (ibid. 495).
[76] 'Doctrine', *TI* xiv. 43 (*ST* x. 282).

nuances. The source of these was twofold: first, his analysis of how the teaching authority arrived at its decisions; secondly, his assessment of the impact of the modern world, particularly of pluralism. In addition to detailing these points, what follows will also concentrate on identifying both what Rahner regarded as the principal challenges to the magisterium in the present, and how it could expect to be reshaped in responding to those challenges.

While reference to the Inquisition or the Holy Office normally evokes images of a shadowy group claiming omniscience and doggedly resisting modernity, it was these very stereotypes which Rahner sought to exorcise in his discussion of the magisterium. Consequently, he argued that any precipitate recourse to anathemas, the traditional weapon of those responsible for maintaining orthodox belief, served only to fossilize the Church's doctrine.[77] Far from encouraging either reaction or repression, Rahner's explorations of the role of the teaching office sought to identify not only its limits, but also its resources for responding positively to the challenges of a changing world. To begin, what was required was a proper understanding of how the office actually functioned.

As interpreted by Rahner, the fundamental feature of the teaching authority was its dependence on the general community of faith—a point which was made in Chapter 2 in regard to the development of doctrine. Both its authority and its knowledge of the truth were derived from that community. Indeed, the articulation of the community's faith was the primary duty of the magisterium. If separated from the faith of the wider Church, the teaching authority was reduced to what Rahner dismissively termed a 'magico-mechanical' approach to the truth.[78] The teaching authority did not, therefore, derive its power directly from Christ, but from the Church as a whole. Consequently, Rahner claimed that there was no mere 'listening' Church which existed simply to be addressed by those in authority.[79] Those who held office, those who exercised the legitimate right to teach, did so only because they too were members of a community of believers. A believing

[77] 'Heresies Today', *TI* xii. 137–8 (*ST* ix. 474–5).

[78] 'Magisterium and Theology' ['Magisterium'], *TI* xviii. 61 ('Lehramt und Theologie', *ST* xiii. 76), orig. pub. 1978.

[79] 'The Teaching Office of the Church in the Present-Day Crisis of Authority' ['Authority'], *TI* xii. 5–6 ('Das kirchliche Lehramt in der heutigen Autoritätskrise', *ST* ix. 341), orig. pub. 1970.

Church—composed both of those who held office and those who did not—was thus a prerequisite for a teaching authority.

The magisterium's necessary dependence on the faith of the wider Church also determined its manner of operation. Thus, those who held this office needed, for example, both to study the relevant theological, historical, and exegetical issues and be open to discussion if they were to maintain a connection with the faith consciousness of the whole Church.[80] Such a process was possible only after laying to rest the fear that divine support for truth in the Church was at odds with the human activity which established that truth. In short, what had to be learnt was that, as had been clearly demonstrated by votes taken at every Council from Jerusalem to Vatican II, democracy and the teaching authority were not mutually exclusive. Significantly, Rahner did not restrict either to bishops or to a Council the right to be involved in consultations prior to the formation of doctrine. While recognizing that contemporary conditions militated against direct democracy, he none the less emphasized the need for some wider discussion in order to ascertain the Church's faith conviction, which, as was acknowledged in an earlier chapter, was more than mere public opinion.[81]

In the process of the teaching authority becoming aware of developments in the Church's faith, theology was a key participant. Furthermore, since theology was itself not static, the possibilities and challenges it presented to the magisterium also varied in every age. Rahner's understanding of how the pluralistic age had affected the relationship between theology and the teaching authority will be the major concern in what follows.

PLURALISM, THEOLOGY, AND THE MAGISTERIUM

As has already been indicated, Rahner believed that, in formulating doctrine, the teaching authority represented the faith conviction of the wider Church. In practice, this meant that the teaching authority gave formal approval to the understanding of faith developed by contemporary theology: 'In its defining or authentic doctrinal statements, the magisterium sanctions a development

[80] Ibid. 11 (ibid. 347). [81] Ibid. 15–21 (ibid. 351–6).

of the Church's sense of faith which had been stimulated and sustained by unofficial theology *before* this sanctioning.'[82]

Ironically, however, it was the magisterium's dependence on theology which convinced Rahner that its ability to issue definitive statements of faith would in future be limited. The modern world, suggested Rahner, had produced so many 'balls'—understood as the differentiation in the sciences, anthropology, history, and other branches of the human sciences—that no juggler could attempt to play with them all.[83] In other words, the pluralism of philosophies, terminology, and horizons of understanding which characterized contemporary theology would make it impossible to produce new infallible statements which could encapsulate the single faith-consciousness of the entire Church.[84]

In addition, Rahner was convinced that pluralism would also affect the future disciplinary role of the magisterium. In an age when both curial congregations and non-curial theologians shared the same *Verstehenshorizont*, it had been relatively easy to identify and censure those who departed from the norms. That age, however, was past. Thus, while *Humani Generis* had been able not only to formulate unequivocally the positions it opposed, but also positively to refute them, Cardinal Ottaviani's 1966 letter to the world's bishops—which was discussed in the previous chapter—could only refer in a general sense to disturbing modern trends.[85] Indeed, contemporary complexity was such that Rahner suggested that adequate supervision of the emerging plethora of theologies required a corresponding plethora of teaching authorities.[86] If this was not to be done, then the magisterium had to risk allowing individual theologians to be responsible for preserving their adherence to the Church's creed.[87] At the very least, the new situation obliged both parties to engage in dialogue.

The need for dialogue was strengthened by the necessarily critical function which was exercised by theology towards defined

[82] 'Magisterium' *TI* viii. 60 (*ST* xiii. 76). Rahner's emphasis.

[83] *Karl Rahner in Dialogue*, trans. H. D. Egan (New York, 1986), 295 (*Karl Rahner im Gespräch*, ii [*Gespräch*], ed. P. Imhof and H. Biallowons (Munich, 1983), 208).

[84] 'On the Concept of Infallibility in Catholic Theology' (Infallibility'), *TI* xiv. 73 ('Zum Begriff der Unfehlbarkeit in der katholischen Theologie', *ST* x. 312), orig. pub. 1970.

[85] 'Philosophy', *TI* ix. 60 (*ST* viii. 83).

[86] Ibid. (ibid.).

[87] 'Creed', *TI* xi. 18 (*ST* ix. 26).

dogma. Rahner stressed that the proper mission of the theologian was not simply to repeat and analyse what was taught by the magisterium, but, out of an evolving social context, to question the Church's confession of faith.[88] Indeed, he prized the theologian's sowing of a 'creative restlessness' within the Church.[89] By raising what was unreconciled with Church teaching, the theologian was to be the catalyst for its reconciliation. Significantly, Rahner believed that this activity was not only to precede magisterial statements, but also to follow them, since even defined teaching had to be made comprehensible in varying intellectual, cultural, and historical situations.[90] Rahner not only commended this activity, but as illustrated by his own response to *Mystici Corporis*, actively engaged in it.

This emphasis on the interpretative function that the theologian exercised in regard to defined dogma expressed Rahner's conviction that the Church's truth was historically relative. In fact, he identified two forms of this historical relativity. The first related to the conceptual models by means of which the truth was expressed; the second to the impact of particular truths in different periods of history, to the fact that not every truth had the same revolutionary force in every age.[91] In what follows, the focus will be mainly on the first point.

The possibility of a history of dogma derived from the necessary dependence of dogmas on human concepts which originated in a particular cultural, religious, and secular environment.[92] Since even their employment in Church teaching did not confer immutability on such concepts, Rahner argued that there could be no 'perennial' doctrine; that is, no doctrine immune from the need for continuing interpretation.[93] Indeed, he recognized a danger that some dogmas might actually become unintelligible as the contemporary meaning of words of 'pre-theological' origin outstripped their usage in the Church's doctrine. Thus, for example, there was an ongoing need to clarify how such terms as 'person' or even

[88] *Grace*, 175 (*Gnade*, 137). [89] Ibid. 176 (ibid. 138).

[90] 'Magisterium', *TI* xviii. 61 (*ST* xiii. 77).

[91] 'Doctrine', *TI* xiv. 41 (*ST* x. 280).

[92] 'Basic Observations on the Subject of Changeable and Unchangeable Factors in the Church' ['Changeable'], *TI* xiv. 9 ('Grundsätzliche Bemerkungen zum Thema: Wandelbares und Unwandelbares in der Kirche', *ST* x. 247), orig. pub. 1972.

[93] Ibid. 9–10 (ibid. 247–8).

'bread and wine' were used in the context of Church teaching, since their secular meaning had developed beyond what was current at the time of Chalcedon and Trent respectively.[94]

That the magisterium's reliance on the concepts available at a particular moment in history was unavoidable meant that there could never be such a thing as 'chemically pure' doctrine, uncontaminated by the spirit of its time.[95] Rahner emphasized that even solemnly defined dogmas were 'amalgams', since they inevitably clothed timeless truths in the philosophical and linguistic dress of a transient age. Thus, Trent's teaching on original sin needed to be separated from its connection to Monogenism, a connection which was sustainable in the sixteenth century, but not in the twentieth.[96] Like a miner of alluvial gold, therefore, the theologian had the task of sifting the sparkle of abiding truth from the mud encasing it.

As the contemporary pluralism of theologies meant that new definitions of doctrine were unlikely in the future, this same pluralism also gave birth to the possibility that even existing dogmas would be interpreted in a multitude of ways. As a result, Rahner suggested that commentaries on dogma—which attempted to express the meaning of particular dogmas for a world no longer sharing the linguistic forms of the dogma—would become more important in the future than the dogmas themselves.[97] At the same time, however, he also affirmed the exclusive right of the magisterium to determine a *Sprachregelung*; that is, to specify both the words to be adhered to, and how language was used in the Church's doctrine.

If the Church was to have unity in its profession of faith, it needed to be unified in its understanding of doctrine, hence the *Sprachregelung*.[98] Although this notion assumed major importance in Rahner's work in the post-conciliar period, it had been adumbrated—as was seen in Chapter 1—as early as his 1947 article on membership of the Church. While the process of specifying the meaning of doctrine had always been part of the Church's

[94] 'The Historicity of Theology', *TI* ix. 71–2 ('Zur Geschichtlichkeit der Theologie', *ST* viii. 96–7), orig. pub. 1966.

[95] 'Changeable', *TI* xiv. 10 (*ST* x. 248).

[96] 'Magisterium', *TI* xviii. 62 (*ST* xiii. 78).

[97] 'Creed', *TI* xi. 15–16 (*ST* ix. 24).

[98] 'Heresies Today', *TI* xii. 132 (*ST* ix. 468–9).

patrimony, Rahner believed that, in the future, it would become the primary activity of the magisterium. He envisaged the heart of the teaching authority's future operations as being neither the formation of new doctrine nor even the further development of ancient teachings, but emphasizing particular points of such teachings in order to defend their truth against false interpretations: 'Precisely in the future, therefore, the Church's decisions on doctrine will appear as (justified and important) regulations of meaning [Sprachregelungen] for the Church's one profession of faith and will be received as such.'[99]

Rahner argued, however, that even the *Sprachregelung* was not immune to the pressures of history. In drawing a line of demarcation between what was regarded as reconcilable with the Creed and what not, the magisterium was necessarily deciding on issues of epistemology and sociology, rather than simply on truth *per se*.[100] Furthermore, since the magisterium itself had no option but to express its teaching by means of particular theological concepts, discussion about whether other theologies fell within the bounds of the *Sprachregelung* would necessarily continue.[101] The mere fact that issues of truth were not as obvious as '2 + 2 = 4' suggested the urgent need for a continuing dialogue between theology and the teaching authority in order to arrive at truth.[102] As will be seen in what follows, Rahner's advocacy of this dialogue did not represent an attempt to curtail the prerogatives of the teaching office, but rather expressed a yearning for the magisterium to become more aware of the impact of pluralism.

THE DIALOGUE OF THEOLOGY AND THE MAGISTERIUM

While defending the legitimate role of the theologian, Rahner firmly rejected the suggestion that theologians constituted a separate magisterium: two bodies exercising the same function would only have imperilled the Church's unity.[103] Not only did he thus reject any possibility of theologians competing with the magisterium

[99] 'Authority', *TI* xii. 23 (*ST* ix. 358).
[100] 'Creed', *TI* xi. 18 (*ST* ix. 27). [101] Ibid. 19 (ibid. 28).
[102] '"Mysterium Ecclesiae"', *TI* xvii. 153 ('"Mysterium Ecclesiae"', *ST* xii. 498), orig. pub. 1973.
[103] 'Magisterium', *TI* xviii. 58 (*ST* xiii. 74). For another point of view on the possibility of theologians exercising a separate magisterium, see Avery Dulles, 'The Two Magisteria: An Interim Reflection', in id., *A Church to Believe in*, 118–32.

for authority in the Church, he also highlighted the limits of theology itself. Accordingly, he reminded theologians that they too inevitably operated with a set of 'amalgams' which meant that their work, no less than that of the teaching authority, was not free of the boundaries imposed by particular philosophies. Since much of the tension between theology and the magisterium was generated by the clash of these often irreconcilable philosophies, rather than by the clash of conflicting truths, Rahner urged that they be the focus of dialogue.[104]

If hope of a fruitful dialogue thus obliged theologians to look to their limitations, the same hope also made demands on the teaching authority. In what follows, the focus will be on two such demands, both of them reflecting the historical situation of the contemporary Church: first, the need to appreciate that authority alone could not guarantee acceptance of Church teaching; second, the need to specify the degree of obligation attached to each of its teachings.

A constant refrain of a series of articles which Rahner published during the 1970s was his plea for the magisterium to become aware that, in order for its teaching to be efficacious in an environment shaped by pluralism, more was required than insistence on uncritical obedience. It was not sufficient, he insisted, that the teaching authority simply made and proclaimed its decisions. What was also necessary in the present age, an age characterized by a 'partial identification' with the Church, was a sensitivity to what would make those decisions effective.[105] Authority alone was not the answer.

Furthermore, since acceptance of the formal authority of the magisterium was not the foundational element of Catholic faith, it was inevitably more endangered than the more fundamental aspects of that faith.[106] In fact, Rahner claimed that acknowledgement of the magisterium as an authoritative source of Christian teaching actually entailed a risk analogous to that involved in the central act of faith itself.[107] Formal authority alone did not furnish

[104] 'Magisterium', *TI* xviii. 66–7 (*ST* xiii. 83–4).

[105] 'The Congregation of the Faith and the Commission of Theologians' ['Congregation'], *TI* xiv. 102 ('Glaubenskongregation und Theologenkommission, *ST* x. 342), orig. pub. 1970.

[106] 'Authority', *TI* xii. 24–5 (*ST* ix. 359).

[107] 'Teaching Office', *TI* xiv. 95 (*ST* x. 335).

a criterion sufficient to justify that risk. Equally, such acceptance could not be granted on the basis that the magisterium was sincere in its conviction: sincerity was not incompatible with error.[108] For contemporary believers, therefore, it could only be the content of a teaching, a content which had to relate to the totality of Christian faith, which grounded the authority of the teacher.[109] This in turn reinforced the need for dialogue with theology, since such dialogue, which would make clear that magisterial decisions were not capricious, could help to establish the credibility of the magisterium.

Since this dialogue was to centre on the concepts and the modes of understanding—the 'amalgams'—employed by the teaching authority, it needed, for the reasons discussed above, to continue even after the proclamation of a doctrine.[110] In promoting such dialogue, Rahner was not conducting a disguised campaign against the rights of the teaching authority. In fact, he acknowledged that the inevitable gap between the general and the concrete meant that no amount of dialogue could substitute for decisions. Nevertheless, he was similarly insistent that decisions alone could not replace dialogue.[111] Furthermore, since effective dialogue needed time, he urged both sides to avoid ending it precipitately by recourse to whatever forms of power they could invoke. Additional support for an open and patient dialogue, Rahner found in *Gaudium et Spes* (*GS* 43), which directed that, in disputes between the laity and the teaching authority, neither side could claim the authority of the Church for its position alone.[112]

So convinced was Rahner of the need for dialogue that he even suggested it be institutionalized, if this could be done without depriving it of the capacity for the unexpected and the improvised.[113] Any attempt at such institutionalization needed, however, to respect that the magisterium retained the *Kompetenz der Kompetenz*.[114] Nevertheless, the complex nature of the relationship between theology and the teaching authority meant that, as in all other aspects of the Church's life where the preservation of unity

[108] Ibid. 96 (ibid. 336). [109] 'Authority', *TI* xii. 27–8 (*ST* ix. 363–4).
[110] 'Magisterium', *TI* xviii. 68 (*ST* xiii. 85).
[111] 'Dialogue in the Church', *TI* x. 111 ('Vom Dialog in der Kirche', *ST* viii. 434), orig. pub. 1967.
[112] Ibid. 112 (ibid. 435). [113] Ibid. 113 (ibid. 436–7).
[114] 'Magisterium', *TI* xviii. 71 (*ST* xiii. 89).

was paramount, nothing could substitute for the activity of the Spirit.[115] Equally, the work of the Spirit did not render superfluous human efforts to promote unity. In this regard, the need for the magisterium to specify the status of each of its decisions, the second of Rahner's prerequisites for a fruitful dialogue, was crucial.

In urging that the teaching authority make clear how binding on the believer was each of its teachings, Rahner was in fact appealing to the magisterium to state publicly that not all of its decrees were as binding as an infallible statement. Such a step, he believed, would not only spare believers the conflicts of conscience which had characterized the reception of *Humanae Vitae*, but would also be a sign of sincerity.[116] Furthermore, Rahner argued that this could be done without implying that the declarations of the teaching authority were either mere theological opinions or entitled to respect only if their arguments merited it.[117] In order properly to appreciate why knowledge of a doctrine's level of obligation was so crucial for Rahner, what has to be considerd again is his notion of the 'amalgams' inseparable from the Church's teaching.

As has already been noted, all decisions of the teaching office necessarily employed the philosophy, theology, science, and linguistics available at any given time. Consequently, there was a danger that further developments in those fields would result in some aspects of Church teaching—like the defence of Monogenism—being shown to be in error. Rahner suggested, therefore, that if all magisterial utterances commanded absolute assent, there would be no openings for further theological research to overcome such weaknesses.[118] This would have meant, for example, that, of the teachings propounded in the last century alone, believers would be obliged to accept that the Pentateuch was written by Moses, the Psalms by David, and that the Index was—necessarily, and unquestionably—a permanent part of the Church's discipline.[119]

In an age of unprecedented social development and mass communication, even constraining the laity and theologians to a *silentium obsequiosum* could not disguise faults in the Church's formulations.[120] In order that the teaching authority might remain

[115] 'Magisterium', *TI* xviii. 63 (*ST* xiii. 79). [116] Ibid. 65–6 (ibid. 82).
[117] Ibid. 65 (ibid.). [118] 'Teaching Office', *TI* xiv. 93 (*ST* x. 333).
[119] Ibid. 92–3 (ibid. 332). [120] Ibid. 94 (ibid. 334).

both credible and trustworthy, Rahner urged that, without waiting for future research to separate the darnel of history from the wheat of truth, it have the courage to acknowledge the errors in what had previously been put forward as the faith of the Church.[121]

In addition, he stressed that the Church had nothing to fear from the fact that its non-infallible teaching might be regarded as provisional. Indeed, echoing the sentiments of a 1967 document of the German Episcopal Conference on the status of non-infallible statements, he claimed that, in a rapidly evolving world, the Church could not be obliged to choose between infallible statements or silence.[122] In such a context, therefore, a provisional teaching, subject to later correction and elaboration, was actually the proper means by which the teaching authority could offer guidance to the faithful. That such statements might contain errors was not to be wondered at by those who appreciated not only the burden of history, but also the inevitable slowness of development of the Church's faith.[123] It was these principles which shaped Rahner's response to *Humanae Vitae*.

Accordingly, he stressed that the encyclical was to be taken seriously, since it was the word of the Pope, was maturely prepared, and had a long Catholic tradition behind it.[124] This meant that it was entitled to the presumption of truth. At the same time, however, he also emphasized that individual Catholics, and particularly theologians, had the right—and even the duty—to be aware of its reformable nature; that is, to be aware that the presumption of truth which it enjoyed might have to be corrected.[125] Although he certainly believed that the encyclical had its shortcomings, Rahner's main concern was not to campaign against it. Indeed, he accepted that *Humanae Vitae* might even be regarded as an 'ideal norm' (*eine 'Zielnorm'*), which, although it could not be realized in the present, none the less represented an important

[121] 'Congregation' *TI* xiv. 106–7 (*ST* x. 347).

[122] The relevant excerpts from the document can be found in Josef Neuner and Heinrich Roos (eds.), *Der Glaube der Kirche in den Urkunden der Lehrverkündigung*, 12th edn., rev. Rahner and Karl-Heinz Weger (Regensburg, 1971), 319–20. Rahner's discussion of the document is in 'On the Encyclical "Humanae Vitae" ' ['Humanae Vitae'], *TI* xi. 267–74 ('Zur Enzyklika "Humanae Vitae" ', *ST* ix. 281–7). This article was orig. pub. 1968.

[123] 'Teaching Office', *TI* xiv. 93 (*ST* x. 333).

[124] 'Humanae Vitae', *TI* xi. 264 (*ST* ix. 277).

[125] Ibid. 271–2 (ibid. 284).

goal for humanity; in this case, it would be similar to monogamy, which, though it was always consistent with human nature, had taken a long time to establish.[126] This positive 'spin' notwithstanding, Rahner also provided guidance for laity and theologians who found difficulties with the encyclical.

As an encouragement for theologians, Rahner noted the progress in magisterial documents from *Humani Generis* to *Lumen Gentium*. While the former specifically vetoed further discussion of even non-defined papal teaching (DS 3885), the latter directed only that such teaching be received with 'religious obedience' (*LG* 25).[127] In regard to *Humanae Vitae*, Rahner interpreted this obedience as implying neither a duty to defend the encyclical as if it were the only possible teaching nor the obligation to keep silent, but rather the obligation to make clear the document's meaning and the reasons for the Pope's position. This obedience was not, therefore, irreconcilable with also indicating the difficulties of such a position and helping people come to a proper formation of conscience.[128] Similarly, Rahner claimed that a believer who, after self-critical reflection, rejected the papal teaching did not need to feel disobedient towards Church authority or constantly to confess such a decision.[129]

Rahner emphasized that the possibility that dissent from non-infallible statements might be abused was not sufficient reason to outlaw such dissent, especially when the alternative would involve both a stagnation of Catholic theology and the prospect of many people falling away from the faith.[130] If post-Vatican II Catholics were not yet able to appreciate their freedom to the extent that they understood the principle of *epikeia*, this bespoke not too much freedom—which might have legitimized restriction—but too little experience with that freedom.[131]

Thus far, the analysis of Rahner's arguments in support of the

[126] 'Humanae Vitae', *TI* xi. 264 (*ST* ix. 277). The ref. to the 'ideal norm' is on p. 273, and the discussion of the encyclical's weaknesses on pp. 277–9 (ibid. 286 and 290–2).
[127] Ibid. 283 (ibid. 296–7). [128] Ibid. 283–4 (ibid. 297–8).
[129] Ibid. 284–5 (ibid. 298–9).
[130] 'Teaching Office', *TI* xiv. 94 (*ST* x. 334).
[131] 'Freedom and Manipulation in the Church', in *Meditations on Freedom and the Spirit* [*Meditations*], trans. R. Ockenden, D. Smith, and C. Bennett (London, 1977), 66 (*Freiheit und Manipulation in Gesellschaft und Kirche* [*Freiheit*] (Munich, 1970), 49).

necessity for dialogue between theologians and the magisterium has concentrated on the application of such dialogue to the Church's reformable teaching. What remains to be explored, however, is Rahner's attitude to infallible teaching.

THE INFALLIBLE FAITH OF THE CHURCH

Rahner's promotion of dialogue between theologians and the teaching authority was characterized by one major 'non-negotiable': theologians were obliged to accept that any doctrine expressly claimed to be absolutely binding for Catholics could not be in error. It was this conviction which, in 1971, triggered a major clash between Rahner and Hans Küng. Since it illustrates much about Rahner's attitude to the tensions between theologians and the teaching authority, this particular clash of the Titans merits closer examination.[132] For Rahner, the two central points of the debate were: Küng's rejection of the error-free status of infallible statements, and Küng's categorization of *Humanae Vitae* as both infallible and faulty. Since the debate involved Rahner responding to Küng, any analysis of it must begin with the latter's thesis.

Briefly summarized, the kernel of Küng's argument was that even defined propositions could contain error. To prove his claim, he concentrated on *Humanae Vitae*, which he regarded as both equivalent to an infallible statement—since it represented an unbroken tradition—and flawed.[133] Küng asserted that *Humanae Vitae*, because it symbolized the magisterium's failure to acknowledge errors in its teaching, was the Achilles' heel of Roman claims to infallibility.[134] As an alternative to an infallible teaching authority, Küng portrayed the Church as a community on pilgrimage to the truth. While the community as a whole could not fall away from truth, this neither implied nor required that Scripture, Councils, or the Pope enjoyed the charism of infallibility.[135]

[132] A contemporary analysis of the dispute can be found in John J. Hughes, 'Infallible? An Inquiry Considered', *TS* 32 (1971), 183–207. A summary of the dispute favourable to Küng—indeed, it accuses Rahner of 'obsequious sabotage' in his response to *Humanae Vitae*—can be found in Robert Nowell's, *A Passion for Truth: Hans Küng—A Biography* (London, 1981) 201–22. Rahner's own reflections on the dispute are given in *Karl Rahner in Dialogue*, 93–100 (*Gespräch*, i. 181–92).

[133] Hans Küng, *Infallible? An Enquiry*, trans. E. Mosbacher (London, 1971), 48–52.

[134] Ibid. 145. [135] Ibid. 147–58.

In reply, Rahner claimed that Küng had contradicted both the general principles which had characterized Catholic theology since the Reformation, and the expressed teaching of both the First and Second Vatican Councils.[136] For that reason, Rahner claimed that he had to respond to Küng as he would to a liberal Protestant with whom Catholic theology could find no common ground.[137] As interpreted by Rahner, Küng's position amounted to reconstructing Catholicism after his own fashion.

Not only did Rahner thus repudiate the general tenor of Küng's position, he also specifically rejected the latter's claim that *Humanae Vitae* could be described as infallible. While accepting that Paul VI had never specifically acknowledged that the encyclical was reformable, Rahner refused to endorse Küng's assertion that this was tantamount to the attribution of infallibility.[138] In addition, Rahner was unmoved by Küng's opinion that the flaws in the encyclical were self-evident. Without retracting his own criticisms, Rahner none the less refused to accept either Kung's dismissal of *Humanae Vitae* or his assertion that the encyclical had been comprehensively rejected by the Catholic community.[139]

Underpinning Rahner's criticism was his conviction that human existence required legal propositions to objectify basic decisions—a further application of the relationship between truth and institution, which was discussed in Chapter 3. Rahner contended that if such propositions exist, they must also exist in the Church, since the Church represented the ultimate decisions of humanity in regard to God's offer of salvation.[140] Consequently, absolute assent to defined dogma was a constitutive element of Catholic faith and, therefore, a prerequisite of Catholic theology. Furthermore,

[136] 'A Critique of Hans Küng: Concerning the Infallibility of Theological Propositions', *Homiletic and Pastoral Review*, 71 (May 1971), 13 ('Kritik an Hans Küng: Zur Frage der Unfehlbarkeit theologischer Sätze', in Rahner (ed.), *Zum Problem Unfehlbarkeit: Antworten auf die Anfrage von Hans Küng* (Freiburg i.B., 1971), 31. This book was ed. by Rahner and included his article on infallibility from *ST* x and his two replies to Küng, of which the above was the first. Küng's replies to Rahner's criticism can be found in *Homiletic and Pastoral Review*, 71 (June and July 1971), 9–29 and 17–50 respectively.

[137] 'A Critique of Hans Küng', 13 (*Zum Problem Unfehlbarkeit*, 32). For further discussion on the notion that dialogue with Küng was like dialogue with a 'liberal Protestant', see Rahner's contribution to Rahner and Küng, 'A "Working Agreement" to Disagree', *America*, 129 (7 July 1973), 11–12.

[138] 'A Critique of Hans Küng', 16 (*Zum Problem Unfehlbarkeit*, 35–6).

[139] Ibid. 14 (ibid. 33). [140] Ibid. 21–2 (ibid. 40–2).

in contrast to Küng, Rahner asserted that such consent was not contingent on a particular dogma being proven to be congruent with Scripture and tradition. Rather, the very proclamation of a doctrine was itself proof of its necessary congruence:

The actual faith consciousness of today's Church is, for me, itself a theological instance, and precisely so, as this expresses itself in the doctrinal decisions demanding an absolute assent of faith of the ordinary and extraordinary teaching office of the Church. According to my understanding of theology, this has always belonged, and still belongs, to the fundamentals of Catholic theology.[141]

The Church alone, therefore, not historical research, determined whether a doctrine was in continuity with the history of faith. While this did not invalidate the need for the interpretation of dogma, Rahner reiterated his conviction—which was discussed in Chapter 1—that the validity of an interpretation depended on its acceptance by the faith consciousness of the whole Church, as represented by the teaching office.[142] Consequently, those who claimed to have found errors in propositions with which the Church had declared itself to be absolutely identified thereby placed themselves outside the one community of faith.

Although Rahner was at pains to defend the legitimacy of infallibility, he was equally concerned that its impact on the life of the believer not be exaggerated. Consequently, he stressed that acceptance of the magisterium's right to proclaim infallible teaching was grounded in one's faith in Christ, rather than in the Church itself. Infallibility did not, therefore, usurp the proper ordering of Christian belief.[143]

Thus, even Vatican I's decree on infallibility, while establishing the infallibility of all preceding dogmatic statements, could not make itself infallible, but relied on being accepted as an expression of Catholic Christian truth.[144] In addition, Rahner emphasized

[141] 'Reply to Hans Küng: In the Form of an Apologia Pro Theologia Sua', *Homiletic and Pastoral Review*, 71 (Aug./Sept. 1971), 15 ('Replik. Bemerkungen zu: Hans Küng, Im Interesse der Sache', ibid. 53). The pub. trans. omits the final sentence of the quote.

[142] 'A Critique of Hans Küng', 16–17 (*Zum Problem Unfehlbarkeit*, 56).

[143] Ibid. 18–19 (ibid. 58–60).

[144] 'Infallibility', *TI* xiv. 76–7 ('Zum Begriff der Unfehlbarkeit in der katholischen Theologie', *ST* x. 315–16). Orig. pub. in 1970—that is, before the dispute with Küng—this article, like Küng's book, was designed to commemorate the centenary of the definition of infallibility.

that infallibility functioned as a relatively secondary instance of control within the life of the Church. Its primary function was to resolve conflict directed towards other secondary aspects of that system, not to establish the legitimacy of the system itself.[145] Another aspect of Rahner's efforts to keep the doctrine of infallibility in perspective was his prognosis of its future role.

In previewing the future, Rahner's basic assertion was that the impact of pluralism was already such that it was no longer possible for any definition to be accompanied by the claim that it expressed the faith consciousness of the whole Church.[146] Any new definition would be subject to such a variety of interpretations that it would be unable to claim that it actually represented an advance in the Church's faith. What would remain to the teaching authority, however, would be the exercise of the *Sprachregelung* in order to maintain the historical continuity of Christian faith.[147]

In the light of Rahner's efforts to highlight both the second-rank status of the doctrine of infallibility, and its likely minor role in the future, it might be wondered why he so swiftly and strongly opposed Küng. If, as Rahner himself had consistently maintained, even infallible statements were immune to neither interpretation nor development, was Küng really asserting anything so revolutionary in rejecting the possibility of an infallible teaching authority? In order to appreciate Rahner's response to Küng, it is necessary to realize that, ironically, it was not infallibility *per se* which was the crucial issue for Rahner: the real issue was Rahner's understanding of what it meant to be a theologian within the Church. This can be clearly seen in a small apologia which appeared during his dispute with Küng:

> I have always felt myself as a theologian 'within the system' and never wanted to be anything else. I have constantly campaigned against any narrowness of this system and any interpretations not worthy of belief; but always in the name of the 'system' itself and by invoking principles which the 'system' recognises as its own.[148]

[145] 'Infallibility', *TI* xiv. 77 ('Zum Begriff der Unfehlbarkeit in der katholischen Theologie', *ST* x. 316). Orig. pub. in 1970—that is, before the dispute with Küng—this article, like Küng's book, was designed to commemorate the centenary of the definition of infallibility.

[146] Ibid. 72 (ibid. 310–11). [147] Ibid. 82 (ibid. 322).

[148] *Homiletic and Pastoral Review*, 71 (Aug./Sept. 1971), 18 (*Unfehlbarkeit*, 58). The pub. trans. gives only an abriged version of this passage. Rahner's profession of his belief that theology belonged within the 'system' earned from Küng the

If Küng's rejection of infallibility was perceived by Rahner as a threat to the ecclesial nature of theology, this alone was sufficient to justify Rahner's opposition. As the above quote illustrates, however, Rahner's ecclesial sense was not irreconcilable with criticizing the Church—indeed, such criticism could be seen as a holy obligation, as it helped to clarify the real nature of the Church. Thus, when Küng's licence to teach was withdrawn in 1980—as a result of rejection of his Christology—Rahner not only expressed the view that the action was without sufficient grounds, but also made clear his opposition to the procedures of the Congregation for the Doctrine of the Faith.[149] In addition, as will be shown in what follows, he did not shrink from some scathing denunciations of the teaching authority's failure adequately to present and defend the Christian faith in the twentieth century.

THE FALLIBLE PRACTICE OF THE MAGISTERIUM

This analysis of Rahner's understanding of the teaching authority's role in the Church began with the claim that he sought to highlight the positive contribution that the magisterium could make to the life of the Church. His reflections on the actual operation of the teaching authority, however, occasioned him little joy. Indeed, Rahner characterized the work of the Congregation for the Doctrine of the Faith during the twentieth century as a *historia calamitatum*, a chronicle of missed opportunities presided over by people who believed that they alone had access to the truth.[150] On this point, Rahner's conviction was strengthened by his personal experience: in 1971 he resigned from the International Theological Commission—to which he had been appointed only in 1969—because he believed that the Congregation took no notice of the Commission.[151]

For Rahner, the proof of the Congregation's failures was provided by the fact that most of its pre-Vatican II statements on the

reproof that such a claim identified Rahner as 'the last great (and stimulating) Neo-Scholastic'; see Küng, *Theology for the Third Millennium*, trans. P. Heinegg (New York, 1988), 188.

[149] For Rahner's statement regarding lack of grounds in regard to the action against Küng, see Norbert Greinacher and Herbert Haag (eds.), *Der Fall Küng* (Munich, 1980), 413–16; for his criticism of the Congregation see, 'Zu: "Warum schweigt Rahner"', *Publik-Forum*, 25 Jan. 1980, 32.

[150] 'Congregation', *TI* xiv. 101 (*ST* x. 341).

[151] *Karl Rahner in Dialogue*, 300–1 (*Gespräch*, ii. 217–18).

Bible had become obsolete and that even the condemnation of Modernism, the high-water mark of its influence, had not dealt adequately with either the mentality or problems which had given birth to that crisis.[152] Rahner's basic accusation was that the Congregation had radically failed to appreciate the dimensions of the contemporary situation. In short, its mind-set, particularly its tendency to interpret the world of today and tomorrow out of the context of yesterday, was an icon of the Church's flawed response to the modern world:

> We take for granted certain norms of behaviour which people of today de facto do not follow. We lay down principles and maxims, which might be very good in themselves, but which unfortunately have the flaw that the people of today simply will not recognise them as their own. We believe that the style in which one is accustomed to think, feel and live in a clerical milieu is also valid outside it.[153]

In Rahner's analysis, the teaching authority had been unable to come to terms with the challenge of pluralism. Consequently, instead of a prophetic approach bringing the Church's ancient faith into dialogue with the questions of the time, instead of making the substance of faith alive in a secularized world, it had limited itself to issues on the margins of the hierarchy of truths.[154] This lack of courage, Rahner found particularly characteristic of *Mysterium Ecclesiae*. This document, which the Congregation for the Doctrine of the Faith issued in 1973 as a refutation of 'certain present-day errors' in ecclesiology, he regarded as being, in spirit if not in letter, prior to Vatican II.[155] Prominent in the litany of Rahner's objections to *Mysterium Ecclesiae* were: its refusal to acknowledge the Church's sinfulness; its failure to appreciate the advances in ecumenism; and its narrow interpretation of the magisterium's exclusive right to teach, an interpretation which disregarded the contribution of theologians and the faith of the Church.[156]

In addition, Rahner was critical of the abstract approach to the history of dogma which was evident in *Mysterium Ecclesiae*. In

[152] 'Congregation', *TI* xiv. 101 (*ST* x. 341–2).
[153] Ibid. 105–6 (ibid. 346). [154] 'Magisterium', *TI* xviii. 72 (*ST* xiii. 91).
[155] The document can be found in Flannery, *Vatican Council II*, ii. 428–40. The original appeared in *AAS* 65 (1973), 396–408.
[156] ' "Mysterium Ecclesiae" ', *TI* xvii. 141–7 (' "Mysterium Ecclesiae" ', *ST* xii. 484–90),

particular, he rejected its assertion that it was easy to distinguish between a dogma's mode of expression and its content.[157] As has already been noted, Rahner was convinced that the amalgams with which the Church's teaching were necessarily incorporated meant that it was no simple matter to distil permanent truth from its time-conditioned expression. Indeed, he claimed that the belief that there actually was continuity in the Church's history of dogma required a decision of faith, as its truth was not susceptible to historical investigation.[158]

In a very idealized portrait in a 1980 article, Rahner envisaged a situation where the teaching authority would acknowledge its errors and commit itself both to more openness and to 'due process' for theologians.[159] Such an approach was consistent with his conviction—which will be the theme of the next chapter—that, as a 'system', the Church had to remain open. In practice, this meant the magisterium welcoming pluralism, the faith instinct of believers, and the new historical situations which gave birth to new questions.[160]

In the late twentieth century, a new challenge for the magisterium was the emergence of a 'world-wide' theology, a theology whose roots were not exclusively European. Rahner recognized the potential problems such a theology might have in dealing with a teaching office—and specifically with the Congregation for the Doctrine of the Faith—whose theology was exclusively European.[161] Without denying that the magisterium had a duty of supervision in regard to these new theologies, Rahner also urged that the teaching office refrain from any 'premature interference' designed to regiment non-Europeans.[162] Instead of a policy of confrontation, Rahner again advocated the virtues of a dialogue which could forestall the emergence of conflict situations.

Despite his disappointment with the concrete performance of the magisterium, Rahner remained steadfastly committed to his basic belief that theologians could not legitimately separate themselves from the faith of the Church or, therefore, from the activity

[157] Ibid. 151 (ibid. 495). [158] 'Changeable', *TI* xiv. 12 (*ST* x. 250).

[159] 'Theology and the Roman Magisterium', *TI* xxii. 176–90 ('Die Theologie und das römische Lehramt', *ST* xvi. 231–48).

[160] 'Humanae Vitae', *TI* xi. 286 (*ST* ix. 300).

[161] 'Aspects of European Theology', *TI* xxi. 86 ('Aspekte europäischer Theologie', *ST* xv. 92), orig. pub. 1983.

[162] Ibid. 87 (ibid. 93).

of the teaching authority. Indeed, in an article published only a year before his death, he asserted that

> if theologians want to live in the Church with the theology they want to live with, then their theology must really be an ecclesial one, and it must have in principle an unprejudiced and positive relationship to the ecclesial teaching office. Those days are certainly gone when a theology could be a Denziger theology. . . . but that is a far cry from saying that the only way young theologians can bring theology alive is, in sovereign ignorance, by no longer condescending to take into account the doctrinal decisions of the Church, and by attempting to pursue theology without formal and continuing dialogue with the Church, its teaching office, and its theological history.[163]

REVIEW

In the previous chapter, it was noted that Rahner regarded Vatican II's openness to the modern world as the most important aspect of the Council. In the present chapter, the emphasis has been on his conviction that such an openness required that the Church not only be aware of the contours of that world, but also develop appropriate responses to it. In Rahner's assessment, the greatest challenge for the Church in the post-conciliar period was to recognize and respond to pluralism.

One of the implications of pluralism, and of the secularization which was attendant on it, was that the religious world-view had ceased to be dominant. Accordingly, Rahner encouraged the Church to concentrate on the essentials of Christian faith, essentials which he believed would continue to resonate with the human spirit even in the midst of conflicting world-views. In the light of this conviction, he claimed that it was a renewed understanding of the fundamentals of faith, rather than a further development of doctrine, which would be the characteristic feature of the future. For so doing, Rahner has been criticized for failing to apply consistently his own pre-conciliar understanding of the development of doctrine, an understanding which emphasized an unfolding of the mystery of God.[164]

[163] 'A Theology that We Can Live with', *TI* xxi. 107–8 ('Eine Theologie, mit der wir leben können', *ST* xv. 112), orig. pub. 1983.

[164] For criticism of Rahner's later approach to development, see Nichols, *From Newman to Congar*, 266–9.

Lest this criticism convict Rahner of inconsistency, it is important to take into account two other principles of his pre-conciliar work on doctrine and its development—features examined in Chapters 2 and 3. The first principle to be recalled is that all doctrinal statements are necessarily expressed in the philosophy and language of their time. This principle suggested not simply the need for continuing interpretation—which Rahner acknowledged would go on even in the context of pluralism—but also the difficulty of development when, as in an age of pluralism, there was no longer unity in either philosophy or language. The second principle is that development might take place as concentration and not simply as extension. In other words, focusing more clearly on the mystery of God, rather than simply gaining new knowledge about God, was the heart of any development of doctrine. Ironically, it was the situation created by pluralism which facilitated such a concentration.

While Rahner took seriously the obstacles to belief created by pluralism, he none the less resisted the urge to tailor the Church after a pattern which might have offered an easier route to popularity. Instead, he insisted that it was the Church's faithfulness to its own fundamentals which was the key to its future. At the same time, however, his desire to see the message of the Church clearly proclaimed meant that he gave short shrift to those in the Church who placed obstacles in the path of such clarity. Hence his emphasis on the need for dialogue between the magisterium and theologians, and his attacks on the lack of openness of those in authority. Indeed, it was openness which Rahner sought more than anything else. Accordingly, the focus of the next chapter will be on his view of how such openness could reshape the Church in the future.

7

The Open Church and the Future

AMONGST the gems of practical wisdom heard by Alice during her sojourn in Wonderland was the King of Hearts' guide to story-telling. His Majesty decreed that a story should begin at the beginning, proceed to the end, and then stop. While the regal injunction is admirably succinct, fulfilling it presents no small challenge. In the present context, it has indeed been possible to examine what might be called the beginning and middle of the story—namely, Rahner's analysis of both the Church's constitution, and the impact of the twentieth century on the Church; detailing the end of the story, however, presents a major difficulty.

Clearly, the 'end' in this situation ought to describe the eschatological state of the Church. Rahner, however, specifically excluded the possibility of human beings obtaining such knowledge. If the Church was truly the sacrament of Christ, the presence in history of God's salvific love, then the specifics of its future—where 'future' refers to the consummation of God's Kingdom—were necessarily as inaccessible as everything else pertaining to the unfathomable mystery of God. The last word of the Church's history could, therefore, be spoken only by God. What then can be done to fulfil the King of Heart's instruction?

Rahner's response to the conundrum concerning the future of the Church combined both due deference to the divine prerogative in regard to the *eschaton*, and a willingness to conjecture about the shape of the Church in the more immediate future. Although he acknowledged that the obligation to profess the *docta ignorantia futuri* was a corollary of confessing that God alone was Lord of history, Rahner's own forecasting did not imply that he had substituted a crystal ball for the more usual tools of theological analysis.[1] Rather, his predictions were obtained by extrapolating

[1] 'The Question of the Future', *TI* xii. 181 ('Die Frage nach der Zukunft', *ST* ix. 519), orig. pub. 1969.

from the impact that the evolving history of the twentieth century had already had on the Church.

As was illustrated in each of the last three chapters, reflection on the changing position of the Church in the modern world dominated Rahner's ecclesiology in the years after the Second World War. In response to these changes, Rahner sought both to identify the resources which would enable the Church to adapt to a rapidly developing world, and to consider how the application of those resources would alter the face of the Church itself. In this chapter, the focus will specifically be on the changes which he envisaged as likely in the following areas: the role of authority in the Church—including the issue of how conflict between office-holders and others might be resolved; the shape of ministry; Christian communities; ecumenism; and the relationship between the Church and the world.

Before examining these issues, however, what needs to be highlighted is the principle which underpinned Rahner's approach to the future: his conviction that the history of the Church could not be regarded as complete, that the final stage of its development had not been reached. In short, Rahner was committed to the idea that the Church was an 'open system' guided by the Spirit. Analysis of Rahner's notion of the open Church can best begin by contrasting it with its antithesis: the closed Church of the Pian period.

THE SPIRIT AND THE OPEN CHURCH

As was discussed in Chapter 4, Rahner believed that the two characteristic features of the Church during the period from Pius IX to Pius XII were: an emphasis on the unchangeable, and an emphasis on the conviction that, in fields as diverse as liturgy and administration, the whole Church ought to mirror what was done in Rome. As a result of this centralizing tendency, the Western Church had little or no experience of differentiation in the years after Vatican I. In addition, the Pian era laid exclusive stress on papal primacy as the source of unity. This alone ensured that anything which smacked of heterogeneity was also suspected of being an attack on the primacy.[2] Against such a background, even

[2] 'On the Theology of a "Pastoral Synod"' ['Pastoral Synod'], *TI* xiv. 118–19 ('Zur Theologie einer "Pastoralsynode"', *ST* x. 360), orig. pub. 1970.

Vatican II's tentative encouragement of the local Church seemed little less than revolutionary. Indeed, Rahner argued that even the accusation that Vatican II had fomented unrest in the otherwise homogeneous history of the Church indicated how deeply the ecclesiology of the Pian period had impressed itself on the Church's self-understanding. Furthermore, in numerous interviews during his last years, he emphasized that opposition to ongoing reform and, its corollary, the desire for a restoration of a pre-conciliar Church, suggested that thraldom to the Pian model persisted even two decades after the Council.[3]

Without denying the reality of disquiet in the post-conciliar Church, Rahner rejected the allegation that the Council itself was to blame. Indeed, he claimed that in addressing itself to what threatened the Church in the present, the Council had performed a task which was so important that it could only have been postponed, not ultimately avoided.[4] The turbulence subsequent to the Council resulted from the presence in the Church of a number of heterogeneous factors which, although they became more evident after Vatican II, did not originate with it. Thus, anxiety before the new existed alongside a craving after novelty; similarly, a loss of faith in the current strategy of the Church was paired with a genuine desire for a better way of life for Christians and the Church.[5] Since the Church's recent history had emphasized homogeneity, there was little readiness to accept either that such tensions were a normal feature of any living organism or that they had actually existed prior to Vatican II. There was, in short, little readiness to accept the need for the Church to remain open.

In such a climate, the challenge for Rahner was to show that the notion of the Church as a closed and centralized community, the model suggested by the Pian epoch, was an aberration. To substantiate this contention, Rahner reiterated one of the key claims of his own pre-conciliar ecclesiology: that the Spirit, the catalyst of change, worked through all the members of the Church.

In 'Observations on the Factor of the Charismatic in the Church',

[3] This point can be found, e.g., in *Karl Rahner in Dialogue*, 278–9 (*Gespräch*, ii. 183–4); and *Faith in a Wintry Season: Conversations and Interviews with Karl Rahner in the Last Years of his Life* [*Faith in a Wintry Season*], ed. H. Egan, 173 and 196–7 (*Glaube in winterlicher Zeit: Gespräche mit Karl Rahner aus den letzen Lebensjahren* [*Zeit*], ed. P. Imhof and H. Biallowons (Düsseldorf, 1986), 211 and 240).

[4] 'Christian Living', *TI* vii. 3 (*ST* vii. 11).

[5] Ibid. 4 (ibid.).

an article written in 1969, Rahner claimed that if God was truly its Lord, the Church could not be considered a closed system whose every aspect could be adequately explained from a point within the system itself.[6] A closed system—a designation suggestive of totalitarianism—would have meant, for example, that all developments in the Church were contingent on the initiative of the Pope and other office-holders, who would have been sole possessors of the wherewithal to facilitate change in such a system. If, however, the impetus for growth was actually provided by the Spirit, whose activity in the Church was not confined to the hierarchy, then impulses for change could be expected to emerge from every group constituting the Church's social reality.[7] Contemporary trends in the Church confirmed for Rahner that this expectation was not vain.

Prominent among modern developments highlighted by Rahner was the growth of 'individualization' in the Church. Not only was he convinced, as has been noted, that in a society where faith and Church membership had ceased to be coterminous with citizenship, only those prepared to take personal responsibility for their faith would in fact join the Church, he also believed that such individuals would want to be listened to, to engage in dialogue with the hierarchy, rather than merely to listen.[8] Consequently, the days of paternalism, of omniscient bishops and supine flocks, were numbered. Furthermore, Rahner suggested that if all members of the Church were gifted by the Spirit, if concern for the Church's welfare was not the exclusive prerogative of its office-holders, then the breakdown of paternalism was actually an essential element of the Church's self-realization, rather than a penance which the hierarchy was to endure with stoic resignation.[9]

That the Spirit breathed through all members of the Church, a tenet obscured during the Pian era, had been reaffirmed at Vatican II.[10] In addition to recognizing the universality of the Spirit's mission, the Council Fathers had, in rejecting Paul VI's proposal that *Lumen Gentium* declare the Pope to be answerable only to

[6] 'Observations on the Factor of the Charismatic in the Church' ['Charismatic'], *TI* xii. 88–9 ('Bemerkungen über das Charismatische in der Kirche', *ST* ix. 422–3).

[7] Ibid. 90 (ibid. 423–4).

[8] 'People of the Church', *TI* xii. 222 (*ST* ix. 562).

[9] Ibid. (ibid.).

[10] See e.g. *LG* 7 and 12.

God, specifically defended the *ius divinum* status of the episcopate against possible papal encroachment.[11] If accepted, Paul VI's motion would have attributed to the Pope an autonomy which, argued Rahner, he could not legitimately claim: 'Even the Pope is not merely the helmsman of the history of the Church, but is himself guided through that history whose actual helmsman does not belong to that history.'[12]

Since the charisms of Church members did not have their source in the Pope, it followed that neither he nor any other single authority could specify what form such charisms should take. Consequently, it was not to be wondered at when these charisms flowered in diverse ways. So far was Rahner from endorsing the Pian stress on uniformity that he even advocated freedom within the Church for groups formed around a particular charismatic personality.[13] Noting that the history of religious orders proved that positive developments in the Church were not dependent on being launched by the hierarchy, Rahner urged the contemporary Church to make room even for those 'sects'—understood as small groups protesting against the 'Church of the masses'—whose existence was a source of friction.[14] Perfect order—which was attainable only when a single authority was both the sole innovator and able to enforce a uniform discipline—was not to be esteemed above the individual's free response to grace. It was variety, not uniformity, which manifested the Church's nature as the sacrament of the unfathomable mystery of God. Similarly, it was variety, not uniformity, which affirmed that the Spirit's movement in the Church was also a mystery.[15]

Lest the above summon forth visions of a Church condemned to chaos by the caprice of the Spirit, it must be stressed that Rahner's defence of the open system did not imply the abolition of obedience. It did, however, imply a particular understanding of obedience. In contrast to its meaning in a totalitarian state—where it becomes perverted to unquestioning acceptance of orders received from above—Rahner claimed that obedience in the Church

[11] 'Charismatic', *TI* xii. 90–1 (*ST* ix. 424–5). This ref., in both the Eng. and Ger. versions also contains a footnote indicating where a detailed discussion of Paul VI's proposal can be found.

[12] Ibid. 93 (ibid. 427).

[13] 'People of the Church', *TI* xii. 221–2 (*ST* ix. 561–2).

[14] Ibid. 223 (ibid. 563).

[15] 'Charismatic', *TI* xii. 93–4 (*ST* ix. 427–8).

involved a charismatic element which could not be manipulated either by those holding office or by those called on to obey. More an art than a science, obedience required a synthesis between responsible personal initiative and critical judgement on the law and the directives of the Church.[16] Thus, obedience too was linked to personal freedom and openness to the Spirit, who alone could preserve the Church from anarchy.

The notion of the open Church applied, however, not only to the fact that the Spirit, rather than any point within the system, was the ultimate source of the Church's life, it also related to the Church's inexhaustible—because Spirit-directed—potential for development. In this regard too, Rahner's approach was the antithesis of the mind-set which had dominated the Pian era; a mind-set which assumed that the apogee of development, in both theology and the Church, had already been reached.[17]

Ironically, the Pian era itself, despite its explicit stress on the unchangeable, provided evidence that the history of the Church was anything but complete. The theology of the time may have portrayed the Church as an oasis of stability in a restless world, but this view was not easily reconcilable with such post-Vatican I developments as the introduction of Church-State concordats, the 'internationalizing' of the college of cardinals, the reform of the fast before Communion, and the promulgation, for the first time, of a universal Code of Canon Law.[18]

While theology during the Pian epoch gave the impression that change in the Church occurred despite the Church's best efforts, Rahner insisted that the Church's reflection on itself ought actually to promote change. To support this claim, he argued that there existed in the Church something akin to natural law: a law which derived from the nature of the Church, from the fact that it symbolized the ultimate unity of the world, humanity and history, the self-communication of God, and the presence in history of the eschatologically victorious grace of Christ.[19] Although much of what was regarded as *ius divinum* was an expression of that law,

[16] Ibid. 97 (ibid. 430–1).

[17] 'Structural Change in the Church of the Future' ['Structural Change'], *TI* xx. 116 ('Strukturwandel der Kirche in der künftigen Gesellschaft', *ST* xiv. 334), orig. pub. 1977.

[18] Ibid. 117 (ibid. 335).

[19] *Vorfragen zu einem ökumenischen Amtsverständnis* [*Vorfragen*] (Freiburg i.B., 1974), 36.

this did not imply that it had been fully expressed or exhausted by either scholastic ecclesiology or specific teachings of the magisterium.[20] Consequently, since it was invariably 'wider, freer, and greater' than what was realized in any particular historical form, the Church's self-understanding was the measure by which the actual performance of the Church could be criticized and the need for change identified.[21] The reference in *Lumen Gentium* to the Church as *semper purificanda* (*LG* 8) Rahner interpreted as highlighting the importance of an enduring will to self-criticism in the Church.

Without a spirit of self-criticism, there was a permanent danger that the Church would succumb to the temptation of institutionalization and be regarded as an end in itself.[22] When this occurred, the Church ceased to change, lost contact with other social realities, and became a conservative power.[23] Self-criticism and reform—both of which will be discussed in detail in later sections— could heal the ills of a Church, understood as all members rather than merely those holding office, which preferred the comfort of the *status quo* to the uncertainty of change, even though it was the latter alone which reflected the unfathomable mystery of its Lord.[24]

Having established that the Church's charismatic dimension committed it to self-criticism and reform, Rahner proceeded to identify the areas where such reform was necessary. As will be seen in the sections which follow, he undertook this task with the fervour of a zealot. Indeed, his reforming ardour left few aspects of ecclesial life untouched. Accordingly, he promoted the cause of innovation, or at least modification, both in the Church's relationship to the world, and in life within the Church itself. The linchpin of his proposals for the latter was the conviction that the Church's existence as an open system required that it demonstrate itself to be the home of freedom. Why this was so will be the theme of the next section.

[20] *Vorfragen zu einem ökumenischen Amtsverstandnis* [*Vorfragen*] (Freiburg i.B., 1974), 37.

[21] 'The Function of the Church as a Critic of Society', *TI* xii. 233 ('Die gesellschaftskritische Funktion der Kirche', *ST* ix. 572–3), orig. pub. 1969.

[22] Ibid. 231 (ibid. 571). [23] Ibid. (ibid.).

[24] 'Opposition in the Church' ('Opposition'), *TI* xvii. 132–3 ('Opposition in der Kirche', *ST* xiii. 474), orig. pub. 1974.

THE CHURCH: SACRAMENT OF FREEDOM

One of the ideas discussed in Chapter 3 was Rahner's notion that freedom was best understood not in sociological terms, but as a gift of grace enabling us to transcend our finitude in order to accept or reject God. In short, freedom was to be regarded as a primary constituent of humanity's relationship with God. The Church's existence as the sacrament of Christ, the symbol of God's offer of life to humanity, meant that it was also the sacrament of this freedom.[25] As such, the Church was commissioned not only to safeguard freedom, but to extend it. Consequently, the Church was obliged to leave people free both to make their own decisions and to take the consequences of those decisions.[26] If the Church had claimed the right to total control of an individual's life, it would have meant transmuting itself into an ideology, a system closed in upon itself, rather than open to the unfathomable mystery of God.

As his reference to the danger of ideology reveals, Rahner was aware not only of the tensions which inevitably exist between freedom and any institution, but also that the Church had no immunity from the tendency of institutions to manipulate freedom for their own sake.[27] Indeed, he found evidence of such manipulation in Rome's failure to emphasize that the respect due to its teaching authority did not abrogate the personal freedom and responsibility of every believer. In failing to make clear this distinction, especially in regard to non-defined teachings, the magisterium infringed a right which Rahner identified as belonging to what was most traditional in the Church.[28]

This infringement was not, however, the only instance of the Church trespassing against freedom. Indeed, Rahner argued that the Church's failure to respect freedom was not even restricted to intra-Church matters. Accordingly, he claimed that the Church's record in regard to respecting the rights and legitimate autonomy of civil society was generally dismal:

[25] *Meditations*, 60 (*Freiheit*, 41).

[26] 'Ideology and Christianity' ('Ideology'), *TI* vi. 56 ('Ideologie und Christentum', *ST* vi. 74), orig. pub. 1965.

[27] See e.g. 'Institution and Freedom', *TI* xiii. 112 ('Institution und Freiheit', *ST* x. 123), orig. pub. 1971.

[28] *Meditations*, 64–5 (*Freiheit*, 46–7).

Only too often, naively or culpably, the Church has certainly been associated with the powerful, presented its message as the opium of the people, glorified in a conservative spirit the existing state of affairs, which was anything but good and free, warned against projects for the future, which could have actually brought more freedom and which, contrary to the Church's concrete wishes and prognoses, often actually did so.[29]

The danger of manipulation and narrowness notwithstanding, it remained true that the Church's social constitution was integral to its existence as the sacrament of Christ. Consequently, the tension between freedom and manipulation in the Church, a tension produced by the ineradicable concupiscence of human nature, could be expected to continue. Such an admission did not indicate, however, that Rahner was merely resigned to accepting the manipulation of freedom. In fact, just as he had argued that the damaging effects of pluralism—which derived from the same concupiscence—could be eliminated only asymptotically, he also stressed that a continuing struggle against the sinful misuse of authority in the Church was imperative.[30] If quietism in the face of abuses was not an option for Rahner, neither was the quixotic notion that a future enlightened generation would achieve the final triumph of freedom: grace alone could preserve the Church from the corruption of an ideological system in which the welfare of the institution demanded that freedom be curtailed.[31]

Although he thus doused hopes that human wisdom alone could inaugurate the millennium of true freedom, Rahner was none the less committed to the importance of human efforts to maximize freedom in the Church. The specific measures which he believed could achieve a new order in the relationship between the individual believer and the authority of the Church will be discussed in what follows. As will be seen, his proposals blended three elements: first, defence of the permanent aspects of the Church's constitution; secondly, a refusal to equate permanence with petrification; thirdly, the conviction that the Church's situation in contemporary society made reform of the prevailing model of authority a necessity.

[29] 'The Church's Responsibility for the Freedom of the Individual' ['Responsibility'], *TI* xx. 56 ('Die Verantwortung der Kirche für die Freiheit des einzelnen', *ST* xiv. 255), orig. pub. 1980.

[30] Ibid. 59–60 (ibid. 259).

[31] 'Ideology', *TI* vi. 58 (*ST* vi. 76).

FREEDOM, AUTHORITY, AND REFORM

In advocating a revised understanding of authority within the Church, Rahner was insistent that this did not amount to calling into question the legitimacy of authority.[32] What it did involve, however, was the need to clarify the basis on which such authority was exercised. In particular, it meant recognizing that not all aspects of, for example, the concrete operation of the episcopal office could be sanctioned and defended under the rubric of *ius divinum*.[33] Among those aspects of the practice of authority which could not claim such protection, none was more objectionable to Rahner than paternalism.

If it was true that the Spirit guided all the members of the Church, then paternalism, where those holding office acted as if they were the sole repositories of all virtue and wisdom, was indefensible. To discourage its propagation, and to emphasize that office in the Church was for service of the community, Rahner recommended that both the Pope and bishops hold office only for a limited period.[34] He feared that until office in the Church was associated with service, bishops, *qua* bishops, would continue to adopt postures incongruous not only with their ministry, but even with their personal convictions: 'The *institutionalised* mentality of bishops is, if one may so say it, feudal, impolite and paternalistic; this does not apply to the individual bishop as a concrete person, in whom such behaviour is simply not noticeable, but this makes the situation worse not better.'[35]

As was discussed in the previous chapter, Rahner was convinced that the manner in which bishops and others exercised their authority often constituted a burden for many believers. Indeed, in 1979, in an angry response to the vetoing by Cardinal Ratzinger, the then Archbishop of Munich, of the appointment of Johann Baptist Metz to the Chair of Fundamental Theology in Munich—despite Metz having the unanimous support of the University's authorities—Rahner even alleged that bishops often abused the trust placed in them by members of the Church: 'The Christian at the base of the Church can often have the bitter

[32] *Meditations*, 67 (*Freiheit*, 50).

[33] 'Aspects of the Episcopal Office' ['Episcopal Office'], *TI* xiv. 190 ('Aspekte des Bischofsamtes', *ST* x. 435), orig. pub. 1972.

[34] *Meditations*, 68 (*Freiheit*, 51).

[35] Ibid. 69 (ibid. 52). Rahner's emphasis.

impression that, objectively seen, his or her unconditional loyalty towards the Church and its office-bearers, a loyalty which comes from faith, has been misused.'[36]

Although the radically flawed theology of the feudal model of episcopacy sufficed to condemn it, its theology was not its only liability: as damning was its lack of congruence with the Church's position in contemporary society. A Church existing in the diaspora, where it was dependent on the personal commitment of each of its members, could not demand absolute obedience to authorities who were beyond the influence of those they ruled. In such a situation, it was necessary that authority be exercised in a way which invited, rather than compelled, obedience. If this was done, it could be expected to change the Church:

> The Church, in its present minority situation, where it is no longer supported by the secular powers of society, can only be the Church and can only continually become the Church by deliberately being the Church from below and, interpreted sociologically, by no longer being a Church which is *set before* its members and *confronts* them. Consequently, the relationship between the base and office must be organised in a considerably different way to what we are normally used to today. This new arrangement will not be without its effects on [the Church's] structures and institutions.[37]

As an example of such possible changes, Rahner suggested both that the relationship between bishops and groups of expert advisers ought to be juridically regulated and that, in specific instances, the latter be given a deliberative, not merely consultative, role.[38] In fact, he was prepared to grant bishops the right to veto the decisions of these experts only when such decisions were a positive threat to faith. In addition to enhancing the 'efficiency, transparency and credibility' of the episcopal office, this innovation would have sounded the death knell of paternalism without abrogating a bishop's power of decision.[39] Not only the relationship between a bishop and his diocese, but also that of the local Church to the universal Church was embraced by Rahner's plans for reform. Thus, he argued that national synods—the embodiment of the local Church—ought to be free to make binding decisions, to

[36] 'Ich protestiere', *Publik-Forum*, 23 (1979), 19.
[37] 'Transformations', *TI* xvii. 179 (*ST* xii. 526). Rahner's emphasis.
[38] 'Episcopal Office', *TI* xiv. 193–4 (*ST* x. 438). [39] Ibid. 194 (ibid.).

be accepted as *ius humanum*, without everything being directed from Rome.[40]

In addition to urging a revised understanding of authority in the Church, Rahner also sought a new approach to the management of conflict in the Church. Here too, the innovative nature of his approach can best be seen by contrasting it with that of the Pian era, whose emphasis on uniformity and papal primacy had actively discouraged dissent.

In analysing Rahner's treatment of pluralism, one of the points emphasized was his conviction that the complexity of contemporary thought made it imperative that Church authorities be willing to engage in dialogue with, rather than automatically censure, those holding apparently unorthodox views. Only such a dialogue made it possible to establish whether certain opinions were in fact incompatible with Christian faith. Similarly, he claimed that opposition in the Church was not to be immediately regarded as a sign of evil intent: radical criticism could, as a result of grace, be united with unreserved trust in the Church.[41] Indeed, consistent with his notion that the Church's self-understanding was always greater than that embodied in particular historical forms, Rahner argued that a critical attitude towards the Church was an essential aspect of a believer's relationship to it.[42] Consequently, it was not only true that it was believers attempting to live their faith within the Church, not outside observers, who were best able to assess the Church's performance, it also followed that even the willingness to criticize could be a sign of faith:

> One can reform the Church only *in* the Church and with an eschatological, and—from the human standpoint—not adequately provable hope, to whose nature it belongs to enter, under apparently bad and hopeless conditions, into a struggle against any establishment in the Church.[43]

In addition to acknowledging that both opposition and criticism could be authentic aspects of ecclesial life, Rahner was also willing to countenance the formation of 'parties' within the Church. Such parties, however, needed to be: self-critical; open to discussion; prepared to accept that they alone did not represent what was genuine in the Church; and committed neither to portraying

[40] *Meditations*, 71 (*Freiheit*, 54). [41] 'Assent', *TI* xii. 148 (*ST* ix. 485).
[42] 'Opposition', *TI* xvii. 129–30 (*ST* xii. 472).
[43] 'Schism', *TI* xii. 107–8 (*ST* ix. 442). Rahner's emphasis.

opponents as enemies nor to claiming that those in office were necessarily the foes of the Spirit.[44] One standard by which the legitimacy of groups within the Church could be gauged was their willingness to encourage members of the hierarchy to activate the potential for hope which belonged to their office. Similarly, he suggested that the building of a community of faith, love, and prayer in the Church could be accomplished, not by those who seized the power of institutions, but by those who depended on the powerlessness of the Spirit and of hope. Again using the history of religious orders as his support, Rahner claimed that the most effective reform of the Church had been accomplished by groups set up to achieve a positive aim, not by formalized opposition groups.[45]

Despite his general affirmation of the value of criticism, Rahner was not prepared to give the critics *carte blanche*. Indeed, as he had done in his dispute with Küng, he categorically rejected those advocates of reform who either denied the Church's tradition and self-understanding or regarded the possibility of creating a schism as a legitimate instrument of policy:

A willingness to create a schism destroys the eschatological hope of the Christian that in the Church of Christ the good willed by God will always in the long run find a place through humility, patience and courage without a rupturing of the historical continuity of the one Church, which is not a ruptured synagogue but the Church of the end-time, after which will come only the eternal kingdom of God.[46]

Even freedom of conscience did not give a person the right to claim space within the Church for every conceivable thought and action. As had also been clear in the clash with Küng, Rahner believed that membership of the Church implied accepting that the Church was a community with a particular self-understanding and programme. Membership did not, therefore, include the right to promote a different self-understanding.[47]

For the same reason, Rahner spurned a trend in the post-Vatican II Catholicism towards transforming the Church into a this-worldly power pursuing secular aims. He argued that if the Church abandoned its vertical, eschatological hope in favour of a

[44] 'Opposition', *TI* xvii. 137–8 (*ST* xii. 480–1).
[45] Ibid. 135–6 (ibid. 478–9).
[46] 'Schism', *TI* xii. 107 (*ST* ix. 442).
[47] *Meditations*, 87–8 (*Toleranz*, 22).

horizontal hope, it would cease to have a role in history: a Church which adopted a secular paradigm of liberation, which repudiated the link between salvation and God, would become merely the superfluous agent of the secular State.[48]

Although he thus opposed any secularizing of the Church, Rahner none the less championed the idea that secular models of conflict resolution could be applied within the Church. Among the possibilities thus included were: the right to public debate on matters of conflict; the best-possible flow of information and the widest-possible expression of the reasons for a particular decision; procedural rules for dealing with conflict resolution; the possibility of appeal to a court of higher instance; a variety of arbitration courts; a clear statement of where, when, and by whom actual decisions were made; the right to see files, especially one's personal file; the right to a defender of one's choice when subject to official procedures within the Church.[49] Such measures did not, however, absolve the parties to an intra-Church dispute from their common obligation to the Christian virtues of understanding, love, humility, and a readiness to make concessions.[50] Since both the sinfulness of the Church and the insuperable differentiation of consciousness among its members—due to historical, cultural, and social factors—ensured that conflict in the Church would continue, the need for virtue also remained.[51]

In promoting the adoption of formal procedures which would be used not only to resolve conflicts in the Church, but also to regulate the relationship between office-holders and the people they served, Rahner's primary concern was to increase the realm of freedom in the Church. He also recognized, however, that such measures could increase the bureaucratization of the Church, but claimed that even this danger did not outweigh the good to be achieved by rules which, if rightly framed, could actually protect against regimentation.[52] Not only could intra-Church benefits be thus attained, but the development of a more creative relationship between freedom and authority could constitute the Church as a prophetic witness for the wider society in its struggles with the

[48] 'Opposition', *TI* xvii. 134 (*ST* xii. 477). See also Avery Dulles, 'Vatican II and the Purpose of the Church', in id., *Reshaping of Catholicism*, 132–53.

[49] *Meditations*, 94 (*Toleranz*, 29).

[50] Ibid. 92 (ibid. 27). [51] Ibid. 83–5 (ibid. 17–19).

[52] 'Episcopal Office', *TI* xiv. 195 (*ST* x. 440).

same issue.[53] For this to be accomplished, however, it was necessary for the Church courageously to surrender models of authority—specifically, the feudal one—which secular society had itself already abandoned.

If the effectiveness of the Church in modern society was contingent on reform within the Church itself, then the need for a new order did not apply merely to authority and discipline. Accordingly, Rahner advocated a fundamental 'rethink' in areas as diverse as ministry, community life, and ecumenism. Integral to this process, which shaped his proposals for the future, was to be a theological analysis of the situation in which the Church had to exercise its mission. The task of undertaking such an analysis, and drawing conclusions from it, belonged to practical theology.

PRACTICAL THEOLOGY

Unlike Shakespeare's Juliet, whose comment on the name of the rose reveals disturbing nominalist tendencies, Rahner argued that in regard to practical theology even the name itself was significant. Most importantly, 'practical theology' was to be distinguished from 'pastoral theology'. The need for such a distinction derived from the connotations of the latter term: 'pastoral theology' was the umbrella term traditionally applied to those activities undertaken by priests for 'the cure of souls'.[54] Accordingly, in the pantheon of theological disciplines, 'pastoral theology' had generally been regarded as merely an appendix to moral and dogmatic theology and, consequently, had dealt mainly with confessional practice.[55] Since 'pastoral theology' thus implied that the Church's pastoral mission involved only priests, Rahner preferred 'practical theology' for its capacity to convey the full scope of pastoral ministry.[56]

[53] 'Responsibility', *TI* xx. 58 (*ST* xiv. 256).

[54] 'Practical Theology and the Social Work of the Church' ['Practical Theology'], *TI* x. 350 ('Praktische Theologie und kirchliche Sozialarbeit', *ST* viii. 668), orig. pub. 1967.

[55] 'The New Claims which Pastoral Theology Makes upon Theology as a Whole' ['Pastoral Theology'], *TI* xi. 116–17 ('Neue Ansprüche der Pastoraltheologie an die Theologie als ganze', *ST* ix. 128), orig. pub. 1969.

[56] At times, even Rahner himself used 'pastoral theology' and 'practical theology' interchangeably. In such instances, however, the traditional meaning of 'pastoral theology' was not intended; see e.g. ibid. 121–2 (ibid. 133–4). For a detailed discussion of the relationship of practical theology to Rahner's entire theological enterprise, see Karl Neumann, *Der Praxisbezug der Theologie bei Karl Rahner* (Freiburg i.B., 1980).

In promoting practical theology, however, Rahner's aim was neither to liberate pastoral questions from their subordinate rank nor to establish a separate discipline for them. More radically, he urged that pastoral issues be recognized as the primary responsibility of all systematic disciplines in theology.[57] The justification for such an aim lay in Rahner's understanding of practical theology; for him, it implied

theological reflection upon the *entire* process by which the Church as a whole brings its own nature to its fullness in the light both of its own nature *and also* of the contemporary situation of the world and the Church today, thought out from a theological point of view. Now those actively involved in this task of bringing the Church as a whole to its fullness include *all* degrees in the Church's hierarchy *and* also the laity, as well as everything which is included in the concept of this self-fulfilment of the Church: not only cult, doctrine and pastoral work in the narrower sense as immediately concerned with the mediation of salvation to the individual, but also, no less essentially, the charitable work of the Church and the whole of its impact in all its aspects upon that which we call the 'world'.[58]

Underlying this definition was Rahner's stress on the role of the Spirit in the Church. Consistent with his sacramental approach, he portrayed the Church as a dynamism, as something happening ever new, rather than as something static.[59] For this reason, the Church could not be said to have an essence independent of its existence in space and time, but was continually realizing itself, through the action of the Spirit, in its unique historical situation. Consequently, practical theology was misrepresented if portrayed as a collection of eternally valid conclusions drawn from a knowledge of the Church's essence.

Properly understood, practical theology was an 'original and autonomous' science which, on the basis of a theological analysis of the present, proclaimed with a prophetic tone what was required of the Church if it was to realize its mission in the world.[60] As such, it was not only more than dogmatic ecclesiology, but,

[57] Ibid. (ibid.).
[58] 'Practical Theology', *TI* x. 350 (*ST* viii. 668). Rahner's emphasis.
[59] 'Practical Theology within the Totality of Theological Disciplines' ['Theological Disciplines'], *TI* ix. 102 ('Die praktische Theologie im Ganzen der theologischen Disziplinen', *ST* viii. 135), orig. pub. 1967.
[60] 'Practical Theology', *TI* x. 351 (*ST* viii. 669).

even while making use of them, it also outstripped the social sciences, futurology, and political theology.[61] In addition, its orientation towards a developing response to the world meant that practical theology necessarily exercised a critical attitude towards the Church's existing strategy.[62] In short, it was a reproach to those maintaining that theology was merely 'art for art's sake'.[63] Rahner's enthusiasm for practical theology was not only consistent with his own fundamental ecclesiology, it was also a measure of his fear that post-conciliar theology would either revert to a dalliance with marginal questions or would content itself with commenting on the Council's themes while ignoring contemporary issues.[64] In his own work, Rahner clearly signalled that his commitment was to responding to the challenges faced by the Church in the present. One of the fruits of this commitment was the *Handbuch der Pastoraltheologie*.

The multi-volume *Handbuch*, for which Rahner was both a joint editor and a major contributor, was far more than a manual for parochial clergy. Produced between 1964 and 1969, the *Handbuch* not only fulfilled Rahner's requirement that practical theology garner the insights of the many branches of theology and those of the secular sciences—the labour of historians, educationists, exegetes, sociologists, moralists, and dogmatic theologians is to be found in the various volumes—it also specifically directed itself towards the situation of the Church in the modern world and was concerned with future possibilities for the Church.[65]

Preparing the Church for the future was also prominent in a project Rahner undertook at the same time as the *Handbuch*: his proposals for reforming the theological formation of students for the priesthood. His ideas can be found in *Zur Reform des Theologiestudiums*, which was published in 1969 as a response to a proposal on formation from the German Bishops' Conference. The programme

[61] 'Practical Theology', *TI* x. 352–3 (*ST* viii. 670).
[62] 'Theological Disciplines', *TI* ix. 104 (*ST* viii. 136).
[63] 'Pastoral Theology', *TI* xi. 122 (*ST* ix. 134).
[64] 'Theological Disciplines', *TI* ix. 110 (*ST* viii. 144).
[65] The four vols. of the *Handbuch der Pastoraltheologie*—vol. ii of which has two sections—were pub. by Herder in Freiburg i.B. between 1964 and 1969; a *Lexicon* followed in 1972. Rahner's contribution to vol. i, on the nature and structures of the Church, was the only section trans. into Eng.; this appeared as *Theology of Pastoral Action*, trans. W. J. O'Hara, (Freiburg i.B., 1968). For the background to the project, see Vorgrimler, *Understanding Karl Rahner*, 83–5.

he suggested was founded on the conviction that any reform must acknowledge the existing pluralism in philosophy and theology, the intellectual background of the students, and the fact that candidates for the priesthood were no longer supported by a homogeneous Christian environment.[66] Accordingly, the centrepiece of Rahner's proposals was for a *Grundkurs* which, beginning with the notion of the human being as the one who questions existence, would introduce new students to the process of reflection on experience, identify God as the centre of that experience, develop the notion of salvation in Christ, and illustrate the relevance of faith to existential questions.[67] In so doing, the course would also aim to show the unity of philosophy and theology.

In regard to the priesthood, however, it was not merely theological education that needed to be reformed. As both the *Handbuch* and Rahner's plan for the *Grundkurs* suggested, it was unlikely that even the shape of priestly ministry could remain unaffected by either contemporary social pressure or the ecclesiology of Vatican II. Indeed, the expansion of pastoral theology to include the ministry of all the baptized was itself a clear indication that a revised understanding of the role and status of the priest was required. Rahner's response to this need was not simply to produce a set of practical suggestions, but to re-explore the very notion of office in the Church in order to determine its potential for development.

In the years following Vatican II, the three characteristic features of Rahner's writings on the priesthood and other forms of office in the Church were: the recognition that the exercise of office had to be connected to the mission of the Church as a whole; the notion that consecration by the Spirit, which was common to the whole of humanity, was the basis of the Church's official ministry; and the claim that although office was essentially one, the Church was free to order it as it chose. The first two principles 'demythologized' clerical office; the third empowered the Church to institute new ministries in response to contemporary needs. As will be seen in the sections which follow, it was these same three principles which framed Rahner's approach to the future of ministry.

[66] *Zur Reform des Theologiestudiums* (Freiburg i.B., 1969), 59–76.

[67] Ibid. 77–96. Rahner's proposals for this *Grundkurs* eventually came to life as *Foundations*.

OFFICE IN THE CHURCH

Even in the years before the Council, Rahner's opposition to any hint of a clerical caste had been unmistakable. Little wonder, therefore, that his post-conciliar writings emphasized not the uniqueness of the clergy, but the relationship of the ordained ministry to the mission of the whole Church. Indeed, lest deacons, priests, and bishops fall prey to the temptation of regarding themselves as a spiritual élite, Rahner stressed that the Church lived not only from office, but from the entire people of God pilgrimaging through history.[68] This meant, therefore, that Christian witness within families and civic communities was often more important than the efforts of office-holders in revealing the Church as the sacramental sign of God's offer of salvation in the world.[69] Consequently, it was inappropriate to portray the priest as set apart from the rest of the Church:

If one understands the office of the priest in relation to the Church, then one can appreciate that it is not in the first instance a power juxtaposed to the Church, understood as the people of God, but as the empowering of an individual in a particular way for the accomplishment of what the Church as a whole is and which must be accomplished in this way through its official ministry, then one must be very careful when asking about the specific ministry proper only to the ministerial priesthood and not to the remaining members of the Church.[70]

Since the exercise of office was not to be separated from the mission of the whole Church, there had to be a link between ordained and non-ordained Christians. This Rahner identified as the consecration which all Christians, indeed all people, enjoyed as a result of God's self-communication.[71] The universality of this consecration meant that priestly ordination did not imply that the new priest was thereby being rescued from paganism. Just as

[68] 'The Future of Christian Communities' ['Christian Communities'], *TI* xxii. 129 ('Über die Zukunft der Gemeinden', *ST* xvi. 172), orig. pub. 1982.

[69] Ibid. (ibid. 172–3).

[70] 'Theological Reflections on the Priestly Image of Today and Tomorrow' ['Priestly Image'], *TI* xii. 46 ('Theologische Reflexionen zum Priesterbild von heute und morgen', *ST* ix. 380), orig. pub. 1969.

[71] This point, which was a consistent feature of Rahner's post-conciliar writings, can be found in articles separated by more than a decade: see both 'Priestly Image', *TI* xii. 43–5 (*ST* ix. 378–80) from 1969, and 'Consecration in the Life and Reflection of the Church' ['Consecration'], *TI* xix. 64, ('Weihe im Leben und in der Reflexion der Kirche', *ST* xiv. 121), which was pub. 1980.

baptism sacramentalized, but did not initiate, God's offer of life to those being baptized, so ordination was not the candidate's first encounter with God. Properly understood, ordination was: 'the historical manifestation and the sociological concretising specification in the dimension of the visible Church of a holiness and consecratedness which has always existed inescapably in that person in the form of an offer in virtue of God's salvific will'.[72]

If all people were recipients of God's self-communication, office in the Church was best understood as the way in which certain Christians lived out their response to God. Thus, the office of a priest or any other ordained minister was neither a task additional to basic Christianity nor a substitute for it, but was, like marriage, a way of being a Christian.[73] The fact that ordained ministers did not have a privileged access to the grace of salvation meant that ordination could not be legitimately portrayed as a promotion in the hierarchy of grace or holiness. Instead, ordination was the conferral by the Church, the basic sacrament, of the grace which would enable the priest to live a life of service to the community; in the marriage ceremony, the Church performed a similar role as God's agent—it expressed God's commitment to the couple pledging their love for one another. It was, therefore, the ministry of the priest which made ordination meaningful, not vice versa.[74] Complementing Rahner's demythologized approach to ordination was his revised understanding of 'the grace of office'.

For Rahner, the 'grace of office' in the Church implied not the ontological transformation of the ordinand, but the promise of God to support the work of those who were ordained.[75] Although such an interpretation fits comfortably with his efforts to relate office to the Church's mission in the world, Rahner never fully resolved whether such an understanding of what was communicated in ordination necessitated a permanent commitment to the priesthood and other offices or whether a temporary priesthood was legitimate. Although Rahner's basic instinct undoubtedly associated priesthood with permanence, his willingness to envisage

[72] 'Consecration', *TI* xix. 67 (*ST* xiv. 124). [73] Ibid. (ibid. 125).

[74] 'Pastoral Ministries and Community Leadership' ['Pastoral Ministries'], *TI* xix. 74–6 ('Pastorale Dienste und Gemeindeleitung', *ST* xiv. 133–5), orig. pub. 1977.

[75] 'On the Diaconate' ['Diaconate'], *TI* xii. 78 ('Über den Diakonat', *ST* ix. 412), orig. pub. 1969.

a temporary priesthood became more pronounced in his later years.

In the immediate post-conciliar period, the permanence of the priesthood had been to the fore as Rahner reiterated the understanding of character which was discussed in Chapter 2. Thus, in an article written in 1969, the need for a permanent priesthood was clearly implied in his view that the office of the priest in making present the word of the Church was so multi-dimensional that it claimed the priest's whole existence.[76] In 1972, however, even while repeating the above claim, he argued that the possibility of a temporary exercise of priesthood could not be rejected absolutely. In this later article, Rahner suggested that if the practice of laicization was taken together with the view that Trent's reference to an 'indelible' character implied not the permanence of office, but that ordination could never be repeated, then a temporary priesthood was conceivable.[77] Indeed, he was convinced that a future increase in laicizations—as a result of social, cultural, and psychological factors—might make this temporary holding of office more common. In addition, he accepted that the existence of temporary vows for those in religious orders might provide a possible model for developments in how the priesthood was to be perceived.[78] Nevertheless, even his final word on the subject—in 1980—did not produce a definitive position. Instead, he argued that although the notion of God's commitment to office-holders made a lifelong character intelligible, the possibility of a temporary priesthood remained open.[79]

Although Rahner thus avoided taking a firm position for or against a temporary priesthood, there were other issues on which his position was less ambiguously expressed. Foremost among them was the third of the three principles identified above as characteristic of his post-conciliar approach to office in the Church: the Church's freedom to order office as it chose.

Fundamental to Rahner's understanding of office in the Church was the belief that it was one. This idea, which was prominent in his work from the late 1960s to the early 1980s, expressed his

[76] 'Priestly Image', *TI* xii. 52 (*ST* ix. 386).

[77] 'How the Priest Should View his Official Ministry' ['Official Ministry'], *TI* xiv. 216 ('Zum Selbstverständnis des Amtspriesters', *ST* x. 463).

[78] Ibid. 217 (ibid. 464).

[79] 'Consecration', *TI* xix. 68 (*ST* xiv. 125).

conviction that office had to reflect the nature of the Church, which was itself one.[80] This oneness did not exclude, however, the possibility of division. Indeed, the various divisions themselves embodied the fact that the unity of the Church was a differentiated unity. Thus, although the one office included the power to teach authoritatively, the power to govern, and sacramental powers, these did not all have to be held by the one person.[81] Consequently, Rahner defended both the Church's right and duty to confer, 'in a variable and graded form', a share in the one office, to impart powers implicit in that office, and to do so sacramentally.[82] In short, he was convinced that:

The Church has an extremely wide-ranging potential to vary her ultimately and necessarily *single* official ministry according to the needs of a particular period and the variety of cultural milieux; to divide this office into various single offices, to regulate the connection between these offices, to fix and to limit the functions of individual office-holders and, without damaging the sacramentality of the conferral of office, to regulate the concrete ways in which office is passed on, perhaps even to distinguish between the usual and what, in cases of necessity, could be the extraordinary ways of handing on of office; and possibly to effect the selection of office-holders through a vote of the laity or some other group entitled to vote.[83]

His advocacy of further divisions of office in the Church did not imply that Rahner was abandoning his belief in the *ius divinum* status of the threefold ministry of bishop, priest, and deacon. He did, however, distinguish between the existence of those ministries, which was *iure divino*, and their content, which could properly be modified by the Church.[84] Once it was accepted that office in the Church was consistent with both 'flexibility and adaptability', the potential for development was great.[85]

Indeed, Rahner suggested that offices could be held not simply by women as well as men, but also by groups rather than merely

[80] Ibid. 69 (ibid. 127). The same point had earlier been made in 'The Point of Departure in Theology for Determining, the Nature of the Priestly Office' ['Priestly Office'], *TI* xii. 34 ('Die theologische Ansatz punkt für die Bestimmung des Wesens des Amtspriestertums', *ST* ix. 368).

[81] 'Consecration', *TI* xix. 68 (*ST* xiv. 125). [82] Ibid. (ibid.).

[83] 'Official Ministry', *TI* xiv. 209–10 (*ST* x. 455–6). Rahner's emphasis is Rahner's.

[84] Ibid. 210 (ibid. 456). [85] 'Diaconate', *TI* xii. 70 (*ST* ix. 404).

by individuals.[86] Accordingly, the designation 'bishop', 'priest', or 'deacon' could be applied in a far broader sense than had traditionally been the case. To support his claim for ongoing development, Rahner argued that the history of the single priestly office revealed that different official functions had at times existed in addition to, and within, the threefold ministry.[87] For the future, therefore, even if the Church chose to restrict the title 'priest' to those who presided at the Eucharist, it was none the less free to broaden the understanding of office in order to confer sacramental ordination on those who, as in the case of teachers, directors of charitable works, psychotherapists, and social critics, exercised a particular role within the community.[88]

The fact that each of the official ministries in the Church could continue to develop also had ecumenical implications. Most importantly, it suggested that the time might come when the forms of ministry which had evolved in other Christian churches might be accepted by Catholics as authentic expressions of innovation in the one Church of Christ.[89] If this was to happen, then a deeper awareness of how developments in the Church's ministry occurred was required.

Rahner argued that the concrete forms in which the Church's single office was expressed did not simply emerge from the realization that the Church was required to preach the Gospel, minister the sacraments, and be concerned with faith, hope, love, and community. Instead, the concrete articulation of office derived from an existential meeting of the office with concrete social conditions.[90] The actual shape of the Church's ministry depended, therefore, on unique historical decisions:

> Praxis (including, therefore, the praxis involved in concrete institutionalisation) is not simply the application of prior general principles, but is rather an event of freedom, and thereby the creation of a unique future which has its own autonomy and is not merely the handmaid of theory.[91]

[86] 'Consecration', *TI* xix. 69 (*ST* xiv. 127).

[87] 'Priestly Office', *TI* xii. 34 (*ST* ix. 369). The sub-diaconate, which existed until the abolition, by Paul VI's Apostolic Letter *Ministeria Quaedam* (15 Aug. 1972), *AAS* 64 (1972), 529–34, of tonsure and of the other minor orders, would be an example of this evolving history of office.

[88] 'Christian Communities', *TI* xxii. 129 (*ST* xvi. 172).

[89] *Vorfragen*, 38–9.

[90] 'Episcopal Office', *TI* xiv. 197 (*ST* x. 442).

[91] Ibid. (ibid.).

So convinced was Rahner of the Church's freedom to be innovative in regard to office that he was even prepared to accept it could introduce new offices without the full support of theology. Consequently, just as the medieval Church had regarded episcopal ordination as sacramental even though the theology of the time had not understood it as such, the modern Church could choose to confer sacramentally 'important and enduring official powers' on those who were *de facto* exercising those powers even without fulfilling the prerequisites for such offices.[92] What Rahner had in mind with such a claim was the possibility of ordaining pastoral assistants.

In a 1977 article on pastoral ministry, Rahner highlighted the anomaly that pastoral assistants were in fact exercising the ministry of a priest without priestly ordination. Pastoral assistants, whose ministry had arisen as a result of a practical need—namely, the shortage of priests—rather than systematic reflection, had gradually taken over all the functions of a priest, except presiding at the Eucharist and the sacrament of Penance, for which sacramental ordination was necessary.[93] As a result, pastoral assistants were already performing a quintessential task of the priest: being the concrete link between members of a particular community and the whole Church. The difficulty for Rahner with this development was not that the role of the priest was thereby being usurped, but that pastoral assistants received no sacramental support for the work they were doing. Rahner argued that a refusal to accept that pastoral assistants were exercising priestly functions reduced the priest to a cultic figure whose patrimony had been eroded to the right to celebrate the Eucharist and Penance.[94] If, however, it was true that pastoral assistants were, *de facto*, doing the work of priests, and if the Church indeed had the power to determine the nature of office, then it was unjustifiable that the Church should make a person leader of a community, but deny them full sacramental empowering for that office.

As early as 1969, Rahner claimed that the key question was not whether there was something which the priest alone could do, but whether it was appropriate that a person performing a complex of tasks which the Church not only judged to be necessary, but which

[92] 'Consecration', *TI* xix. 71 (*ST* xiv. 129).

[93] 'Pastoral Ministries', *TI* xix. 77–8 (*ST* xiv. 136–7).

[94] Ibid. 80 (ibid. 139).

were done in its name, ought to receive the particular grace—that is, via ordination—to aid their exercise of that ministry.[95] A decade later, he questioned whether the refusal to ordain pastoral assistants, who were actually exercising a ministry which was important to the Church's reality, did not imply that the Church was hindering their participation in God's promise to the Church.[96] Furthermore, lest there be any doubt about the appropriateness of ordaining pastoral assistants, Rahner emphasized that their existing ministry was not merely a commission from the Church, but expressed that basic consecration—to which reference has already been made—which derived from God's self-communication.[97] Since priests and deacons did not have a consecration in addition to their office, but because of it, it was appropriate that there should be a sacramental rite for pastoral ministers.[98]

Rahner was in no doubt that the single greatest obstacle to the introduction of a sacramental rite for pastoral ministers was celibacy. He refused to accept, however, that this was an insuperable obstacle. Thus, in his 1977 article on ministry, while acknowledging the value of celibacy, he stressed that the Church's obligation to provide pastoral care, an obligation stemming directly from God, had priority over the demand for a celibate clergy.[99] In 1980, Rahner went even further when, in an article on consecration, he nailed his colours to the mast with the claim that to prize celibacy and academic formation, both in themselves good, above the community's right to a leader, spawned a 'schizophrenia' and 'tacit Protestantization' of the Church, since lay people were already *de facto* leaders of communities.[100] For Rahner, it was anomalous that, from the time of Pius X, the conditions for receiving the Eucharist had been reduced, while the conditions for celebrating the Eucharist had been increased to a 'European maximum', a development which he claimed was incomprehensible to the majority of people in the world.[101]

Rahner's guiding principle was that if leadership of a community and the celebration of the Eucharist belonged together, then lay leaders ought to be ordained and sacramentally recognized for

[95] 'Priestly Image', *TI* xii. 47 (*ST* ix. 381–2).
[96] 'Pastoral Ministries', *TI* xix. 80 (*ST* xiv. 140).
[97] 'Consecration', *TI* xix. 70 (*ST* xiv. 128).
[98] Ibid. (ibid. 128–9).
[99] 'Pastoral Ministries', *TI* xix. 85 (*ST* xiv. 145).
[100] 'Consecration', *TI* xix. 71 (*ST* xiv. 129–30).
[101] Ibid. (ibid. 130).

what they in fact were.[102] Furthermore, he argued that the possibility of ordaining community leaders ought to be dependent not on how many would in fact be ordained, but solely on the right of a particular local community to a leader and celebrant of the Eucharist—a view consistent with his idea of 'relative ordination', which will be discussed shortly.[103] This right remained in force even if the ordination of pastoral assistants did not prove to be a solution for the whole Church.

In advocating sacramental ordination for pastoral assistants, Rahner was not unaware that such a move would produce a different understanding of priesthood from the one which had traditionally held sway in the Church. Rather than fearing such a development, however, he accepted it as consistent with both the Church's freedom to develop office as it chose, and the nature of experiment in the Church.

Rahner emphasized that experiments in the Church, which became necessary when the old models no longer functioned adequately, had an existential character whose results could not be known in advance. Unlike experiments in the natural sciences, which, because the operation of the various factors involved could be predicted, served to confirm or challenge hypotheses, experiments in the Church were affected by the imponderables of practical reason and freedom.[104] Experiments could, therefore, change the Church. Nevertheless, a willingness to experiment was not only consistent with the Church's existence in a world characterized by planning and futurology, it was also a way of witnessing to faith in the Lord of history who could not be manipulated. Indeed, Rahner regarded this willingness to experiment as a constitutive element of the Church: 'Church history is the most radical experiment: where it is not, where it becomes nervous traditionalism, it may perhaps still be the history of sinful people in the Church, but no longer of the Church as it ought to be according to the will of Jesus.'[105]

Accordingly, in the *Handbuch*, Rahner asserted that in the contemporary Church—even if not necessarily for all periods of history—a willingness to take risks ought to be regarded as an

[102] Ibid. (ibid.). [103] 'Pastoral Ministries', *TI* xix. 85 (*ST* xiv. 146).

[104] *Opportunities for Faith* [*Opportunities*], trans. E. Quinn (London, 1974), 215–16 (*Chancen des Glaubens* [*Chancen*] (Freiburg i.B., 1971), 240).

[105] Ibid. 218 (ibid. 242).

imperative.[106] Taking risks did not imply an uncritical conformity to everything new, but the courage to make decisions, decisions which could alter traditional ways of acting, in response to new conditions.[107] Consequently, the question for the Church in the modern world, claimed Rahner, ought to be, not how far was it obliged to go in responding to new situations, but how far could it go 'in exploiting all theological and pastoral possibilities'.[108] Rahner was convinced that if the Church was prepared to act in this way, it would discover that, within the bounds formed by the truth that was its foundation, it had far more freedom for innovation than was often imagined.[109]

Although the concrete future of the Church depended on the unpredictable interplay of historical conditions and specific choices, the ramifications of contemporary social development and the impact of Vatican II on intra-Church life provided a clear indication that even the shape of office in the Church would not remain unaffected. Accordingly, Rahner's proposals for the future shape of the priesthood, the diaconate, and the episcopate envisaged major shifts in direction. As will be seen in the following section, he believed that the biggest single influence on the future shape of office would be the Church's existence in the diaspora and the subsequent development of small communities as the standard form of ecclesial life.

THE FUTURE OF OFFICE IN THE CHURCH

As has been noted with each reference to the diaspora in the course of this book, Rahner believed that the demise of the Church's privileged place in society meant that the future of the Church had become dependent on the personal faith of its members. In a world where membership of the Church would no longer be a corollary of citizenship, the Church would need to be built up from below. Rahner argued that the ideal way to achieve this was via basic communities. Such communities needed not only to be religious groups, but to establish a sense of trust and love between people, a sense lacking in the modern mass society.[110] Through

[106] *Handbuch*, ii/1. 274. [107] Ibid. 275. [108] Ibid. [109] Ibid. 276.

[110] 'South American Base Communities in a European Church', *TI* xxii. 153–4 ('Südamerikanische Basisgemeinden in einer europäischen Kirche', *ST* xvi. 204, orig. pub. 1981.

their proclamation of the Gospel, these communities were to be the focal point for establishing the credibility of the Church in its solidarity with the marginalized.[111] Consequently, their ability to win one person from the contemporary secularized world, not their capacity to keep ten from the remnants of a traditional Christian society, was to be the criterion by which they were assessed.[112]

Since such communities would not necessarily be composed of people from the same neighbourhood, the territorial parish would no longer be the basic unit of the Church. Although Rahner did envisage that existing parishes could themselves become living basic communities, he also argued that non-parochial communities which integrated all aspects of Christian life had a right to recognition from Church authorities.[113] In addition, since basic communities needed to grow from below, they could neither be initiated nor directed by the official Church. They were not, therefore, to be merely an administrative extension of a diocese.[114] Nevertheless, the communities had a responsibility to maintain unity with the wider Church, presided over by the bishop.[115] The nature of the relationship between the bishop and the communities, as well as that of the priest to the community, could, however, be expected to be very different from the feudal model.

Without denying that the future shape of the episcopal office would have to maintain continuity with older forms, Rahner expected that the combination of a willingness to experiment and theological reflection would bring about new styles of dioceses and develop the role of the bishop. One possibility envisaged by Rahner was that the practice of the early Church, where those leading small communities were regarded as bishops, might be revived. If such a development occurred, the college of parish priests—the leaders of communities—would become a college of bishops; the tiny dioceses which would thereby result Rahner regarded as no different from the existing situation of hundreds of Italian bishops responsible for what were really only large parishes.[116]

[111] 'Episcopal Office', *TI* xiv. 199 (*ST* x. 444). [112] Ibid. 200 (ibid. 445).

[113] 'Basic Communities', *TI* xix 162 ('Basisgemeinden', *ST* xiv. 268), orig. pub. 1980.

[114] 'Episcopal Office', *TI* xiv. 199 (*ST* x. 444).

[115] 'Basic Communities', *TI* xix. 162–3 (*ST* xiv. 269).

[116] 'Episcopal Office', *TI* xiv. 200 (*ST* x. 445).

Those who presided over such a college—who would actually be the equivalent of a modern-day bishop—would then be regarded as metropolitans. These 'supreme bishops' were not, however, simply to be administrators, but were themselves to preside over a community. Indeed, Rahner claimed that only those who had displayed the ability to win converts from the secularized milieu ought to be considered candidates for office.[117] Consequently, the bishop would be reliant more on the strength of a charismatic personality than on deference given to his office. These major changes in the form of the episcopal ministry were paralleled by what Rahner believed were likely developments in the shape of the priesthood.

Despite the fact that the image of the priesthood had been relatively stable during the Pian period, Rahner argued that the Church's existence as a historical entity made changes in the form of the priesthood inevitable.[118] He not only accepted this inevitability, but was also convinced that it did not threaten the place of the priesthood in the Church. One possible change was that the priesthood might cease to be regarded as analogous to a secular profession.

Rahner recognized that since the time of Constantine, the priesthood had acquired a social prestige and source of income largely independent of the personality of its holder.[119] This situation was not, however, *ius divinum*. In fact, he asserted that it was feasible that the priesthood could continue to be regarded as a vocation without being viewed as a profession. In other words, it was conceivable that priests might cease to be professionals in a secular sense, might exercise their priesthood on a part-time basis, rather than live 'from the altar'.[120]

In addition, just as the effectiveness of a bishop in a secularized environment was to be measured by his evangelical zeal, so too the success of the priest would depend on a living, religious personality. Consequently, the priest could not be content with simply being an administrator:

> The pastor should clearly not simply be the functionary of an ecclesiastical ritualism; on the contrary, he from his position as leader of the

[117] 'Episcopal Office', *TI* xiv. 201 (*ST* x. 445).
[118] 'Priestly Image', *TI* xii. 39 (*ST* ix. 373).
[119] Ibid. 41 (ibid. 374–5). [120] Ibid. 57 (ibid. 391).

community, should feel a responsibility to witness in an exemplary, living and radical way to his brothers and sisters through his own life.[121]

If the future of the Church depended on living communities of faith, it was clear that the future of the priesthood was not to be separated from such communities. Accordingly, Rahner argued that the style of priesthood which had become dominant in the centuries after Trent, a style which assumed the priest was a civil servant lacking an organic bond to a community—'a mobile state-official who is moved, promoted, acts as a representative of a state which confronts a particular group of human beings as an alien factor armed with power and who alone "organises" them'—had to be jettisoned.[122] To replace it, Rahner urged that each community have the right to nominate its own candidate for ordination.[123] This proposal, which he believed was consistent with both the New Testament and the doctrinal nature of priesthood, Rahner combined with his most challenging idea on priesthood: the possibility of 'relative ordination'.

Relative ordination implied that a priest could be ordained to exercise a leadership role as an artist, journalist, psychotherapist, or specialist in building Christian community, rather than taking on the whole spectrum of priestly work as normally understood.[124] As with the issue of ordaining pastoral assistants, the important question for Rahner was not whether the leadership of a particular group actually required ordination, but whether it was appropriate that those undertaking a leadership role ought to be given sacramental support. This question Rahner consistently answered in the affirmative. His willingness to envisage such major changes in the structure of office did not, however, imply abandonment of his belief that the threefold ministry existed in the Church *iure divino*. To appreciate how the two beliefs could be reconciled, what must be chartered is the development in Rahner's view of the relationship between *ius divinum* and *ius humanum*.

In 1970, in an article written to defend the right of a diocesan synod to be more than an advisory body, Rahner stressed that

[121] 'Christian Communities', *TI* xxii. 127 (*ST* xvi. 169).

[122] *The Shape of the Church to Come* [*Shape*], trans. E. Quinn (London, 1974), 110 (*Strukturwandel der Kirche als Aufgabe und Chance* [*Strukturwandel*] (Freiburg i.B., 1989), 132), orig. pub. 1972.

[123] 'Official Ministry', *TI* xiv. 215–16 (*ST* x. 462).

[124] Ibid. 213–14 (ibid. 460).

human law in the Church was far more than an irrelevant addendum to an unchangeable divine law.[125] In fact, human law was to be understood as the concretization of *ius divinum*, without which the latter would not be real. Whereas he had previously associated *ius divinum* far more with the concrete provisions of the Church's constitution, Rahner now argued that, properly perceived, *ius divinum* was not a concrete practical law, but concerned the continuity of law in the Church, a continuity which Catholics regarded as constitutive of the Church.[126] As an illustration of how *ius humanum* could influence even what was regarded as divine law for the Church, Rahner referred to the election of the Pope. Thus, although it could be argued that continuity in the Church's leadership meant that having a Pope was divine law, the manner of his election was not dependent on the Pope himself—indeed, at the time of each election there is no Pope—and could be altered by the Church.[127] The Church was free, therefore, to determine how the divine law of continuity was to be observed.

Rahner did not just defend the legitimacy of altering *ius humanum*, he even claimed that the obligation to preserve the Church as a community of free belief, hope, and love in a particular age could make the reform of some human laws in the Church—his specific contemporary example concerned the right of the laity to be involved in the decisions of bishops—not merely an absolute moral demand, but a demand of *ius divinum* itself; the implication being that only reform could guarantee continuity.[128] A further area where this freedom to alter human law impinged on the future of office in the Church was in regard to the possibility of ordaining women to the priesthood.

In October 1977, the Sacred Congregation for the Doctrine of the Faith issued *Inter Insigniores*, a declaration which rejected the ordination of women as an option for the Church.[129] The document based itself largely on the 'fact of tradition': that women had always and everywhere been excluded from the priesthood. While Rahner was generally critical of the Congregation's argument, he took particular exception to its use of tradition. For Rahner, reference to an unbroken tradition was not equal to invoking an

[125] 'Pastoral Synod', *TI* xiv. 125–6 (*ST* x. 367–8).
[126] Ibid. 126 (ibid. 368). [127] Ibid. (ibid.). [128] Ibid. 128 (ibid. 370).
[129] The document can be found in Flannery, *Vatican Council II*, ii. 331–45; the original is in *AAS* 69 (1977), 98–116.

absolutely definitive and binding tradition, which would have had to be the product of revelation.[130] He argued that there could be unbroken traditions which were merely human and which, therefore, did not necessarily incarnate truth. Consequently, the practice of not ordaining women could be regarded as a human tradition, which, through social and cultural changes, could become obsolete.[131]

While Rahner accepted both the authoritative nature of the Congregation's statement and the fact that it clearly indicated that the Church had not yet reached the point where a general change of consciousness in regard to women's ordination had developed, he was none the less convinced that the question could not be regarded as closed.[132] Although he accepted that theologians continuing to raise such issues might find the Church's authorities unresponsive—or even hostile—he was convinced that the need to protect the open nature of the Church made it imperative that there be further discussion of women's ordination and other contentious issues.[133]

When compared with the major developments he thus envisaged in the future shape of the episcopacy and the priesthood, Rahner's prognostications for the future of the diaconate, the third tier of the threefold ministry, were minimal. In 1969, in his first major reflection on the diaconate in the post-conciliar period, he acknowledged a change in his own pre-Vatican II view of the diaconate. Before the Council, he had advocated the renewal of the diaconate solely on the grounds that, without ordination, people were actually exercising the traditional ministry of the deacon and, therefore, ought to be ordained.[134] After the Council, however, he broadened this approach by supporting the ordination of those discharging any ministry which was diaconal in nature, even if not the traditional ministry of the deacon; among such new ministries he included that of forming community.[135]

A decade later, when he turned his attention to the future of the diaconate, Rahner's ideas had changed again. As will be clear

[130] 'Women and the Priesthood', *TI* xx. 37–8 ('Priestertum der Frau', *ST* xiv. 211), orig. pub. 1977.

[131] Ibid. 45 (ibid. 220).

[132] Ibid. 38 (ibid. 212).

[133] For his arguments in support of the need to continue discussing controversial questions—in this case, celibacy—see *Opportunities*, 204–9 (*Chancen*, 229–33).

[134] 'Diaconate', *TI* xii. 69–70 (*ST* ix. 403–4).

[135] Ibid. 74–5 (ibid. 408–9).

from the discussion on the future of the priesthood, Rahner's advocacy of ordination for those exercising special ministries within a community implied ordination to the priesthood, not the diaconate. Indeed, he recognized that it had become difficult to define the diaconate in a way which did not suggest that the deacon was simply a minor priest.[136] In addition, if deacons were leaders of communities, then, like pastoral assistants, they ought to be ordained to the priesthood so that they could celebrate the Eucharist. So shadowy had the ministry of the deacon become that Rahner could do no more than acknowledge a lack of agreement among theologians as to its material content.[137] The future of the diaconate depended, therefore, on the Church's freedom to mould its ministry, which, as was noted at the very beginning of this discussion of office, was always essentially one, in response to future historical and pastoral exigencies.[138]

Rahner's conviction that the Church's existence in the diaspora held the key to its future applied not only to the shape of ministry, but also to what became the most controversial of all his post-Vatican II prognoses: the future of ecumenism. As will be outlined in the following section, Rahner's commitment to Church unity, which had been documented throughout this work, issued in his last years in a concrete proposal for the reunion of the various Christian churches, a proposal which was designed to facilitate the preaching of the Gospel in a world where Christians were an ever-decreasing minority.

ECUMENISM AND THE FUTURE OF THE CHURCH

As was discussed in the previous chapter, Rahner believed that in making the attainment of faith more difficult, the pluralist age had ensured that those who nevertheless came to faith would concentrate on the essentials of Christianity. Consequently, without being specifically repudiated, the peripheral aspects of Christian belief would not be prominent in the consciousness of the believer. Paradoxically, this development had the positive side-effect of enhancing the prospects of ecumenism.

Central to Rahner's ecumenical theology in the decade after

[136] 'Pastoral Ministries', *TI* xix. 76 (*ST* xiv. 135).
[137] Ibid. (ibid.). [138] Ibid. 76–7 (ibid. 136).

Vatican II was the conviction that contemporary Christians—and potential Christians—had little or no interest in the dogmatic disputes of the Reformation era. A world dominated by scepticism and rationalism, a world which regarded even today's knowledge as provisional, could not comprehend the traditional subject-matter of ecumenical dialogues: debates over conflicting absolute statements.[139] If there was to be a future for Christianity in such a context, the churches were required to address the questions of their contemporaries rather than continue to debate among themselves issues which non-Christians often regarded as mere semantic quibbles. Rahner was convinced not only that this would mean concentrating on the essentials of Christian faith, but that such an emphasis would draw the churches together:

> The greatest chance for ecumenical theology will consist in working for the Christian theology of the future, when the adherents of all theologies and churches, each in their own way, have learned to bear witness in a credible and comprehensible manner to the Gospel of Jesus Christ as addressed to the person of that future age which has already commenced, no longer demanding of that person as a condition for being a believer that he or she should grapple with unnecessary difficulties in addition to that of metanoia involving the whole person, which must be embraced . . . then, as I hope, these theologies of the future as upheld in the separated churches will increasingly approximate to one another and draw closer to one another than the traditional theologies which are immediately conditioned by the controversial questions of the past.[140]

In making this claim, Rahner was not naïvely ignoring the tenacity of the historical disputes between the churches; he was, however, stressing that a narrow concentration on the issues of the sixteenth century itself ignored the twentieth century's challenges to faith. Since the contemporary context was not that of the Reformation, nor even of Vatican I, modern theology could not focus on the disputes of those times without thereby separating itself from its own historical and sociological context.[141]

For Rahner, proof that the Reformation era ought not to be

[139] 'Is Church Unity Dogmatically Possible?', *TI* xvii. 206 ('Ist Kircheneinigung dogmatisch möglich?', *ST* xii. 558), orig. pub. 1973.

[140] 'On the Theology of Ecumenical Discussion', *TI* xi. 61 ('Zur Theologie des ökumenischen Gesprächs', *ST* ix. 72), orig. pub. 1968.

[141] 'Unity of the Church—Unity of Mankind' ['Unity of the Church'], *TI* xx. 168 ('Einheit der Kirche—Einheit der Menschheit', *ST* xiv. 400), orig. pub. 1978.

allowed to determine contemporary emphases was provided by the fact that the possibility of belief in God, in revelation, and in Christ, all of which were the great issues of the present, had not been disputed by any side in the Reformation controversies.[142] In addition, the impact of pluralism had been midwife to the demise of yet another feature common to all parties in the Reformation disputes: a homogeneity in speech and thought. Shared philosophical and linguistic models had made the debates in the sixteenth century a family matter; in the twentieth, however, the universal comprehensibility of 'God-talk' could no longer be assumed.[143] Consequently, in a world where 'God' was in danger of becoming an alien notion, the traditional points of division—such as the supremacy of the Pope or the manner of justification—could not be given priority over the need to discover ways of speaking about the transcendent. Rahner found support for this opinion in the fact that Vatican II's reference to the 'hierarchy of truths'—which, as has already been indicated, was crucial to his understanding of modern religiosity—had itself been made in the document on ecumenism (*UR* 11).

By the beginning of the 1980s Rahner's conviction that evangelization in a secularized world necessitated a common Christian proclamation of the Gospel had begun to develop into a definite proposal for reunion of the divided churches. Underpinning this proposal was his belief that those theological differences which the churches had yet to resolve among themselves were not sufficient to justify their continued division.[144] The basis of unity was to be found in a common baptism, the shared gift of grace and Scripture, the common celebration of the death of the Lord, and the joint struggle for freedom and justice.[145] Although a united Church did not imply the death of either pluralism or tensions between those from different backgrounds, Rahner argued that for those trusting in God's power at work in the Church, these things alone did not justify delaying moves to unity. In addition, he believed

[142] 'The One Church and the Many Churches' ['One Church'], *TI* xvii. 192 ('Die eine Kirche und die vielen Kirchen', *ST* xii. 542), orig. pub. 1968.

[143] 'Realistic Possibility of a Unification in Faith?', *TI* xxii. 69 ('Realistische Möglichkeit der Glaubenseinigung?', *ST* xvi. 96), orig. pub. 1983.

[144] 'Ecumenical Togetherness Today' ['Togetherness'], *TI* xxii. 86–7 ('Ökumenisches Miteinander heute', *ST* xvi. 118), orig. pub. 1980.

[145] Ibid. 92–3 (ibid. 126).

that what remained of the traditional points of controversy would be raised in a more relaxed manner in the future if Christians were already seeking to identify together the 'innermost centre' of Christian faith.[146]

Rahner's conviction that the remaining differences no longer justified the division of the Churches was the basis of his most radical suggestion for progress towards Church unity. In 1983, in *Unity of the Churches: An Actual Possibility*, a book he co-authored with Heinrich Fries, Rahner claimed that the present level of agreement between the churches was such that unity was possible provided only that no church rejected what was dogma for another, and that no dogma, beyond the Apostles' Creed and the doctrines of Nicaea and Constantinople, was imposed by one church on another.[147] In so arguing, Rahner was extending to the relationship between the churches his understanding of the nature of personal faith in a pluralist age.

Thus, just as he had refused to make membership of the Church conditional on understanding and accepting all its dogmas, so too he claimed that the unity between the churches was not threatened when one church did not fully comprehend the doctrine of another or regarded this doctrine as either existentially irrelevant or unable to be proved.[148] As in the case of the individual, what was crucial was that no church rejected such doctrine as absolutely irreconcilable with the Gospel. Consequently, it was conceivable that some Christian churches, no less than individual Catholics, might have difficulty with the teaching office as a doctrine, but still accept its specific teachings. What was to apply in the one Church of the future in terms of unity of faith was, therefore, to be no different from the conditions which already prevailed within the Catholic Church itself.[149]

If dogma thus provided no insuperable obstacles to unity, what did remain to be overcome was human inertia. If this was to be done, a basic requirement was that the leaders of the separated churches accept that reunion of the one Church was principally a question for them, that there was a need for official action to

[146] 'One Church', *TI* xvii. 193 (*ST* xii. 543).

[147] *Unity of the Churches: An Actual Possibility*, trans. R. C. L. Gritsch and E. W. Gritsch New York, 1985), 25 (*Einigung der Kirchen: Reale Möglichkeit* (Freiburg i.B., 1987), 35), orig. pub. 1983.

[148] Ibid. 33–4 (ibid. 44–5). [149] Ibid. 39 (ibid. 50).

make credible the will to unity.[150] To facilitate the path to unity, Rahner pleaded with the Roman Curia to:

show bold resolution and dare to hope to achieve something of which the end result cannot be calculated in advance, thus displaying itself in its ministries of teaching and leadership in a way demanded by the whole historical situation today; let it eliminate many features of a centralist and bureaucratically administered state seeking to decree from above more or less everything that is at all important; let the limits of the universal primacy as they arise from dogma or can be restricted by the papacy itself be more clearly defined.[151]

In addition, Rahner argued that the Pope needed to avoid giving the impression that unity meant that the Protestant churches would simply be 'taken over'.[152] Just as the Pope had often renounced some of his prerogatives in order to form political concordats, Rahner was convinced that no less ought to be done to support the establishment of a united Church.[153] While the Pope thus had a particular role to play, Rahner urged the leaders of all the churches to devise solutions to concrete pastoral problems such as inter-communion, joint religious instruction, and the Sunday obligation in the context of mixed marriage.[154] In stressing the need for concrete action, Rahner reminded these leaders that their responsibility was not simply to the past, but to the future. Only actions born out of such a conviction could witness to the fact that the truths of faith were not merely to be professed, but also to be lived.[155]

While Rahner encouraged experiments in ecumenical co-operation—even though the will for such experiments was often lacking on the Catholic side—he unequivocally rejected the possibility of forming a 'third Church' by attempting to combine the best features of Protestantism and Catholicism. Far from equating such a notion with Elysium, Rahner dismissed it as no more than a return to post-Enlightenment individualism and the pestilence of

[150] 'Concrete Official Steps towards Unification' ['Official Steps'], *TI* xxii. 80–1 ('Konkrete offizielle Schritte auf eine Einigung hin?', *ST* xvi. 110–11), originally pub. 1982.

[151] 'Unity of the Church', *TI* xx. 171 (*ST* xiv. 402–3).

[152] 'Official Steps', *TI* xxii. 81 (*ST* xvi. 111).

[153] 'Church Union', *TI* xvii. 213 (*ST* xii. 565–6).

[154] 'Official Steps', *TI* xxii. 80 (*ST* xvi. 110).

[155] 'Togetherness', *TI* xxii. 86–7 (*ST* xvi. 118).

sects.[156] True unity presumed the development of an existing—albeit flawed—history of faith, not its denial. Consequently, Rahner urged ecumenical communities not to separate themselves from their 'mother' churches:

> Even if it is difficult, ecumenical basic communities must share the historical burden of the old churches, maintain and cherish a living union of integration and trust with the authorities in the churches; at once humble and proud, they may perhaps regard themselves as the vanguard of the Church on pilgrimage through history, but not as an army that can be seen more closely to be fighting only for its own victories.[157]

The fact that any future united Church would have to maintain continuity with its history was a clear indication that what was being sought was unity not uniformity. Accordingly, Rahner asserted that this particularly challenged Catholics to develop an understanding of unity other than one which envisaged the reconversion of Protestants to a pre-Reformation faith.[158] Thus, he re-emphasized his long-held belief, discussed in Chapter 1, that the history of Protestantism could not be dismissed as 'religiously negative'. Rahner's fervent promotion of unity coexisted with the conviction that a united Church would indeed be in some sense 'Catholic'. He stressed, however, that it would be 'the Catholic Church of the future', formed by the understanding of Christian truth present in the united Church, rather than a carbon-copy of the present.[159]

The need for courage and the willingness to take risks belonged not only to the future of ecumenism, but also to the future relationship of the Church to the world. In this relationship, as also in regard to ecumenism, the Church was challenged to move beyond its history of hostility and develop a more creative response.

A CHURCH OPEN TO THE WORLD

As has already been indicated, Rahner was anxious that the Church of the future define its task primarily in terms of the promotion of the Kingdom of God, rather than of social reform. At the same

[156] 'Third Church?', *TI* xvii. 221–3 ('Dritte Konfession?', *ST* xii. 575–7), orig. pub. 1973, but rev. for its inclusion in *ST* xii in 1975.

[157] 'Basic Communities', *TI* xix. 165 (*ST* xiv. 272).

[158] 'Togetherness' *TI* xxii. 88–90 (*ST* xvi. 120–2).

[159] 'One Church', *TI* xvii. 195 (*ST* xii, 545).

time, however, he also recognized that the unity of the love of God and of neighbour meant that the Church could never allow itself to become merely 'an extremely select department store of holiness in which the individual obtains moral strengthening for the wellbeing of his or her personal soul'.[160] There was, therefore, a need not only to build communities which broke down such spiritual isolation, but for these communities to be aware of realities in the wider world. In the diaspora situation, the Church could not continue

> to act as an impregnable fortress with small arrow-slits in the walls from which the defenders shoot at their enemies. Now it is rather the spacious house with large windows from which one looks out on all spheres of humanity, all of which are encompassed by the creative power and compassion of God. And the altar that stands in the midst of this house is consecrated to God as a sign of grace, not only for those who actually live in the house, but for all who, in a spirit of good will and good conscience, seek to find the way to fulfil their existence.[161]

In short, the Church at all levels was to share, in the opening words of *Gaudium et Spes*, 'the hope and joy, the grief and anguish of the people of our time' (*GS* 1). This was to be done, however, without the Church claiming the right to direct everything. Accordingly, the legitimate 'worldliness of the world' was to be affirmed. Thus, Rahner stressed that Christianity's belief in God as the absolute future did not have as a corollary the belief that only Christians could serve the earthly future.[162] What Rahner hoped had passed was the situation where, because of a lack of respect by Christians for the legitimacy of secular planning, various 'historico-ecclesiastical forms' had become obstacles to genuine human progress.[163]

In rejecting the notion that the Church alone could take responsibility for the future of the world, Rahner was not, however, seeking to banish the Church from the world. In fact, he stressed the crucial role to be performed by the Church in pointing to injustices in modern social development. Since sin had not only

[160] 'New Image', *TI* x. 26 (*ST* viii, 351).

[161] 'The Christian in his World', *TI* vii. 96 ('Der Christ in seiner Umwelt', *ST* vii. 98–9), orig. pub. 1965.

[162] 'Marxist Utopia and the Christian Future of Man' ['Marxist Utopia'], *TI* vi. 65 ('Marxistische Utopie und Christliche Zukunft des Menschen', *ST* vi. 77–8).

[163] Ibid. (ibid.).

a private but also a communal dimension, the Christian could not uncritically accept that social conditions were always and necessarily good.[164] As a result, the Church and individual Christians were to be involved in working for greater social justice. In addition, Rahner suggested that the basic communities, because they established the possibility of personal encounter, could provide a remedy to the depersonalization and institutionalization which had become endemic in modern industrial societies.[165]

One consequence of the Church's commitment to seeking a more just world was likely to be further change in the Church itself. This was so, suggested Rahner, because the Church needed to change in the direction of becoming more just if it was to preach justice to the world.[166] The possibility of such change was another sign that the future of the Church remained open. Indeed, as will be shown in what follows, Rahner believed that it was this openness which ought to remain characteristic of the Church until God closed its history.

THE OPEN FUTURE OF THE OPEN CHURCH

At the centre of Rahner's approach to the future was the belief that the last word in the history of humanity would be spoken by 'that infinite, incomprehensible reality, exalted above all coming to be and perishing, which we call God'.[167] That this was so, that God was 'the absolute future', provided the key to the Church's future.[168] Most importantly, it meant that the Church, the sacrament of Christ, not only had a future but that its future could be completed only by God. Until the *eschaton*, therefore, the Church both continued to exist in history and remained subject to the conditions of history. Pre-eminent among these conditions were: the unpredictability of history, and the need to move forward.

Rahner argued that since God alone was Lord of history, human planning, although in itself good, would never be able to accomplish everything nor would human beings foresee all the

[164] *Shape*, 124 (*Strukturwandel*, 147). See also Jon Sobrino, 'Karl Rahner and Liberation Theology', *Theology Digest*, 32 (1985), 257–60.

[165] *Shape*, 128–9 (*Strukturwandel*, 153–4). [166] Ibid. 127 (ibid. 151).

[167] 'The Future of the Church and the Church of the Future', *TI* xx. 113 ('Die Zukunft der Kirche und die Kirche der Zukunft', *ST* xiv. 331), orig. pub. 1980.

[168] Ibid. (ibid.). Rahner's understanding of the 'absolute future' is clearly expressed in 'Marxist Utopia', *TI* vi. 62–3 (*ST* vi. 81–2).

possible results of their actions.[169] For this reason, the Church's witness to the supremacy of God rather than of human achievement constituted it as the sacrament of 'the unplanned future', the future that belonged to God alone.[170] This did not mean, however, that the Church itself ought to have nothing to do with planning. In fact, urging the Church to plan its future was a consistent feature of Rahner's post-conciliar ecclesiology.

As early as 1968, Rahner identified planning as an expression of hope, as only those who planned had something which they could leave to the Lord of history to decide. The alternative was to abandon hope in favour of 'a mere apathetic sinking into the obviousness of the present'.[171] Planning also revealed the richness of the Spirit's activity in the Church, since it required not merely the application of general Christian principles, but a charismatically inspired creative imagination.[172] One aspect of this creativity was the willingness to make changes before they became absolutely necessary.

Rahner acknowledged that when such changes involved structural reform—as with his own proposals on ecumenism or the future of office—conflicts in the Church were inevitable.[173] While he urged tolerance and openness from both progressives and conservatives, Rahner did not regard the avoidance of conflict as the ultimate aim of the Gospel. Nor did he believe that the *status quo* was always the surest guide to truth—'there is no Christian principle to the effect that the conservatives must always be in the right when a choice has to be made between the two groups [conservatives and progressives]'.[174] Far from suggesting that resistance to change was always to be encouraged, Rahner stressed that the Church's existence in history not only meant that the Church could never separate itself from its past, but also that its future was open and unpredictable, as history was always going forward.[175] Every 'present' was, therefore, a period of transition.

[169] 'Perspectives for Pastoral Theology in the Future', *TI* xxii. 111–12 ('Perspektiven der Pastoral in der Zukunft', *ST* xvi. 150), orig. pub. 1981.

[170] Ibid. 112 (ibid. 151).

[171] 'Perspectives for the Future of the Church', *TI* xii. 203 ('Perspektiven für die Zukunft der Kirche', *ST* ix. 541).

[172] *Shape*, 47 (*Strukturwandel*, 62).

[173] 'Structural Change', *TI* xx. 130 (*ST* xiv. 352).

[174] *Shape*, 49 (*Strukturwandel*, 64).

[175] 'The Church's Redemptive Historical Provenance from the Death and Resurrection of Jesus', *TI* xix. 36 ('Heilsgeschichtliche Herkunft der Kirche von Tod und Auferstehung Jesu', *ST* xiv. 88), orig. pub. 1980.

As has already been remarked in earlier sections, Rahner's commitment to change was not a disguised millenarianism. In fact, in a lecture delivered only a few months before his death, he attacked as naïve both a 'euphoric belief in progress', and the conviction that all in the world would be well if only people listened to the Church—attitudes he regarded as characteristic of *Gaudium et Spes* and of the contemporary Church generally.[176] In the same lecture, Rahner defended the legitimacy of a realistic pessimism which did not expect a paradise on earth. This pessimism had its roots in a recognition of both the contradictory nature of humanity, and of the fact that the Church would not cease to be a Church of sinners.[177] While justified, this pessimism was not, however, to exclude an acknowledgement of God's saving grace.

In his last years, Rahner's reference to the 'wintry season' of reaction and lack of courage in the contemporary Church clearly embodied the pessimism about which he spoke.[178] It was, however, a genuinely Christian pessimism in that, without denying hard truths, it was tempered by hope. That Rahner had not simply resigned himself to being one of the Church's 'angry old men' is clear from the whimsical tone of the letter he wrote from 'Pope Paul VII' to his old friend 'Peppino'.[179] In that letter, Rahner's genuine affection for the Church is combined with his staunch commitment to reforming the Church.[180] Ultimately, therefore, Rahner continued to affirm that even in the future the Church would remain important for those seeking to be disciples of Christ:

> The ecclesial aspect of the spirituality of the future will be less trumphalist than formerly. But attachment to the Church will also in the future be an absolutely necessary criterion for genuine spirituality: patience with the Church's form of a servant in the future also is an indispensable way into God's freedom, since, by not following this way, we shall eventually get no further than our own arbitrary opinions and the uncertainties of our own life selfishly caught up in itself.[181]

[176] 'Christian Pessimism', *TI* xxii. 157–8 ('Christlicher Pessimismus', *ST* xvi. 208–9), orig. pub. 1984.

[177] Ibid. 158–9 (ibid. 210–11).

[178] See e.g. *Faith in a Wintry Season*, 189–200 (*Zeit*, 232–45).

[179] The ref. to the 'angry old men' is Rahner's description of himself; see, *Karl Rahner in Dialogue* 330 (*Gespräch*, ii. 261).

[180] The letter to Peppino appears as the article 'The Perennial Actuality of the Papacy' *TI* xxii. 191–207 ('Die unvergängliche Aktualität des Papsttums', *ST* xvi. 249–70), first pub. 1983.

[181] 'The Spirituality of the Church of the Future', *TI* xx. 153 ('Elemente der Spiritualität in der Kirche der Zukunft', *ST* xiv. 381), orig. pub. 1980.

REVIEW

In Chapter 5, in detailing Rahner's response to Vatican II, it was claimed that his post-conciliar theology could best be understood as the product of a dialectic between the unchangeable divine law and the changeable human law in the Church. This principle is consistently illustrated in his proposals for the future of the Church. In addition, as was pointed out above, his later writings also adopted a broader understanding of *ius divinum* than had been present earlier. Divine law was now understood more as continuity than as content; human law it was which added the content. The catalyst setting in motion this dialectic between human and divine law was, as it had been from the 1940s, Rahner's consideration of the Church's situation in the modern world.

In fact, Rahner's ecclesiology from the end of the Second World War onwards can legitimately be portrayed as an endeavour to come to terms with the Church's existence in the diaspora. Accordingly, his proposals for the future were designed to meet the needs of a Church whose position in the world had changed irretrievably from that of the dominant influence on the prevailing culture to that of the 'little flock'—an expression which is to be understood more as a reflection of the Church's loss of influence than simply as a decline in numbers. What Rahner sought, what he had always sought, was that the Church respond courageously to its new situation. Hence his encouragement both of pastoral theology and of a socio-critical attitude within the Church. Hence too, his urging that the laity in the Church, whose support was crucial to the Church's well-being, be afforded a greater say in shaping its future. This last suggestion has not, however, been universally accepted.

Thus, Avery Dulles argues that any proposal for determining ecclesiastical policy or doctrine 'from below': 'would almost certainly result in an adulteration of the Gospel and would prevent the Church from being able to challenge the reigning secular mentality'.[182]

While Dulles's criticism itself runs the risk of implying not only that bishops are immune to the impact of contemporary attitudes,

[182] 'Ecclesial Futurology: Moving Towards the 1990s', *Proceedings of the Canon Law Society of America*, 47 (1986), 11.

but that they alone have access to such an immunity—a claim that Rahner consistently rejected—his criticism also suffers from not properly representing Rahner's view. Thus, Rahner was not advocating panels of laity usurping the role of bishops—indeed, he specifically affirmed the exclusive right of the magisterium to define the Church's doctrine, an affirmation fully compatible with his stress on the relationship between truth and institution. Moreover, no less than Dulles, Rahner accepted that the Spirit alone could be the source of the Church's strength.[183] What Rahner was championing, however, was the need to give to the charisms of the laity—charisms which applied even in regard to the furthering of the Church's truth—greater prominence and respect than they had traditionally received, especially during the Pian era.

Dulles also rejects Rahner's idea of relative ordination, on the grounds that it would not only weaken the unity of the Catholic community, which depends on universally recognized ministries, but that it would undermine leadership in the Church, whose leaders had always been carefully trained and selected for their reliability as bearers of tradition.[184] While Rahner certainly recognized that his suggestion would alter the traditional understanding of priesthood—alter, but not corrupt—it was this very fact that was encompassed by his approach to the interplay of divine and human law: the Church was not free to abolish the priesthood, but it was free to develop it. Equally, his understanding of the nature of experiments in the Church did not allow him to determine in advance the implications of this new understanding of priesthood; none the less he remained convinced that the Church not only had the freedom to introduce such an innovation, but could be an appropriate response to the Spirit's movement in the present.

More than any other aspect of his approach to the future, however, it has been Rahner's work with Heinrich Fries in *Unity of the Churches* which has been most widely criticized. Indeed, some critics—notably Cardinal Ratzinger, who described the Rahner-Fries plan as 'a theological acrobatic trick, which unfortunately does not stand up to reality', and Aidan Nichols, who wrote of its 'pastoral imprudence' and 'practical irresponsibility

[183] For Dulles on the role of the Spirit, see ibid. 14. [184] Ibid. 11.

vis à vis Catholic doctrine'—have treated the book's theses as an obstacle to genuine progress in ecumenism.[185] It is ironic that much of the criticism has accused the authors of a failure to respect what was in fact most precious for Rahner: the integrity of the Church's faith. The irony is compounded by the fact that Rahner and Fries have been made to appear as advocates of the very syncretism which Rahner had opposed in rejecting the 'third Church'.

In Rahner's defence, it must be acknowledged that the concrete suggestions for reunion were not a spontaneous aberration on his part, but the logical development of his life-long approach to ecumenism. As has been shown, Rahner always accepted the religious value of non-Catholic churches and had combined this with a defence of the Catholic claims to truth. In addition, however, he had always situated ecumenism in the broader context of the Church's position in the world. Ecumenism for Rahner was always an aspect of the Church's obligation to proclaim Christ to the world, an obligation whose fulfilment he placed above the disputes between the churches—especially because he believed such disputes had lost most of their legitimacy. Far from adopting a cavalier approach to the Church's faith, Rahner believed that the future well-being of that faith depended on a united witness in a pluralist world. Ultimately, it was this consideration which determined his proposals for reunion. Even if Rahner's ideas are rejected as being merely a dream, it remains a dream which challenges.[186]

In addition to the rejection of the theory underpinning many of Rahner's suggestions, his proposals for the future have won little support on the practical level. Thus, as part of his analysis of Rahner's approach to the future, Avery Dulles records the indisputable fact that, in many areas, the history of the Church in recent years has not gone in the directions which Rahner suggested as

[185] Joseph Ratzinger, 'Luther und die Einheit der Kirchen', *Internationale katholische Zeitschrift*, 12 (1983), 573; Aidan Nichols, ' "Einigung der Kirchen": An Ecumenical Controversy', *One in Christ*, 21 (1985), 166. For more positive, if still critical, assessments of the book, see the various articles by Arthur Vogel, P. L'Hullier, and M. Fahey in *Ecumenical Trends*, 14 (1985), 97–102. Fries himself documents, and responds to, much of the criticism in *Einigung der Kirchen* (1987 edn.), 157–89.

[186] For Rahner's dream in regard to ecumenism, see 'Dream of the Church', *TI* xx. 133–42 ('Der Traum von der Kirche', *ST* xiv. 355–67), orig. pub. 1978.

likely.[187] Where does this conclusion leave Rahner's proposals for the future and for the open Church in general? Ought he, for example, to have devoted more energy to enriching the prevailing ideas of priesthood, rather than developing proposals such as that for the ordination of pastoral assistants?

While Rahner's theology may have won wider acceptance had his approach been different, settling for such a conclusion fails to do justice to what he in fact achieved. In proposing a model of the future Church, Rahner was interpreting Christian tradition for a new age. This being so, his proposals drew on an aspect of that tradition which is often neglected: its compatibility with development. What Rahner offered was a scheme for developing that tradition in the light of contemporary conditions. While his particular emphases are certainly open to debate, the richness of his insight none the less shines through and continues to offer the Church new possibilities, possibilities which, thus far, have certainly not been definitively rejected.

[187] See e.g. 'Ecclesial Futurology', 11–13.

CONCLUSION

In the months between the Iraqi invasion of Kuwait in August 1990 and the expiry of the ultimatum demanding the withdrawal of all Iraqi troops by 15 January 1991, the morality of a possible war was the subject of extensive controversy. In Catholic circles, the heart of that controversy concerned the notion of the 'just war'. Paradoxically, the inconclusive nature of that debate made at least one thing abundantly clear: that the principles associated with the 'just war' theory could not provide untroubled certainty about the legitimacy, or otherwise, of war in the Gulf. No abstract theory could resolve such imponderables as the Palestinian question, the possible environmental catastrophe which might result from a war, and the danger of further complicating the situation of the hostages in Lebanon. Such issues, however, needed to be aired before the decision for or against war could be considered fully informed.

Similarly, an understanding of change in the Church which concentrated solely on formal principles designed to be applicable in every age and all situations would be of limited value in determining how to respond to particular historical circumstances. Accordingly, what has been emphasized in this work is that Rahner's understanding of change was not simply a universal theory, but a response to the conditions of the second half of the twentieth century. Consequently, although it is possible to isolate some general principles of Rahner's approach to change, his work can be fully appreciated only when it remains anchored in its context.

As was emphasized in the first three chapters, the theoretical elements of Rahner's understanding of change derived from his stress on the sacramentality of the Church. Thus, the Church's existence as a sacrament meant that scope for change was provided not only by the requirement that the Church's 'public face'—its institutions, liturgy, and formal teaching—reflect the God whose commitment to humanity it symbolized, but also by the indwelling

Spirit who served as a permanent catalyst of change in the Church. As has also been shown, however, it was the particular contours of the twentieth century which Rahner regarded as providing the most potent stimuli for change in the Church.

Yet even if Rahner's approach to change in the Church was not merely a discussion of a priori principles, that alone does not establish the value of either his analysis of the processes of change or his specific suggestions for the future. Accordingly, in order to determine whether he can in fact be judged to have contributed positively to the Church's self-understanding in regard to the dynamics of change, the following four 'tests' will be applied to Rahner's ecclesiology: first, it will be asked whether Rahner's approach was internally consistent, whether it all fitted together in the context of his own theology; secondly, since even internal consistency is not incompatible with objective error, Rahner's approach needs to be judged by an external standard—in the present case this will be done by assessing whether Rahner's proposals for change are reconcilable with 'catholicity'; thirdly, the worth of Rahner's contribution can be gauged by the extent to which it indicated the areas where further theological reflection was required—as will be seen, this relates mainly to issues raised but not exhausted by Rahner; fourthly, Rahner's ecclesiology can also be evaluated in terms of its capacity to resolve the tension between conservatives and progressives which, as was discussed in the Introduction, has been a feature of contemporary Catholicism since Vatican II.

In regard to its internal consistency, Rahner's work deserves to score highly—especially if consistency is understood as meaning not that for fifty years Rahner simply repeated the one idea, but that his later work developed in a way reconcilable with what was present, even if only implicitly, in the early writings. In other words, consistency suggests that the process of unfolding which Rahner associated with the development of doctrine can be applied to his own work. Thus, for example, although the contribution those outside the teaching office could make both to the development of the Church's truth and its life in general was more prominent in his later than his early theology, the seeds of this emphasis had always been present in his stress on the consecration and guidance of the Spirit enjoyed by all members of the Church.

Often this consistency revealed itself in what Rahner avoided saying as much as in what he actually said. Thus, although Rahner has been criticized for failing to envisage the possibility of a Church without bishops, this criticism—although true in fact—does less than justice to Rahner's approach to *ius divinum*.[1] Neither in his early work, when he tended to associate *ius divinum* with the content of particular structures, nor even later, when he linked *ius divinum* more with the continuity rather than the content of law in the Church, could Rahner have conceived of a Church without bishops, as this would have involved rupturing the continuity he valued. As we have seen, however, this did not mean that he was opposed to developments in understanding how the episcopal office was to be exercised. Indeed, his uncompromising rejection of the 'feudal' model, as well as his idea that 'bishop' could be used to designate a group rather than merely an individual, were clear applications of the interplay between human and divine law in the Church which was at the heart of his post-Vatican II ecclesiology.

Similarly, to accuse Rahner of failing to follow his theological method to its ultimate conclusion is also to misunderstand that method. Thus, for example, the suggestion that a more rigorous application of the distinction between the 'categorial' and the 'transcendental' would have enabled Rahner to avoid absolutist claims about Christ and the Church in favour of a 'more relativistic Christian position' does less than justice either to the purpose of that method or to Rahner's understanding of truth.[2] The purpose of Rahner's method was not to construct a god, but to illustrate that the central truths of Christian revelation could be confirmed from universal human experience. Consequently, the heart of his method was the conviction that human questioning and self-transcendence have an end, that it is not the fate of humanity to be burdened with an absolute relativism. Thus, while Rahner respected other sources of truth, he did so without denying what for him was the central truth. Indeed, his notion of the 'anonymous Christian' can best be interpreted as an attempt to understand other truths in relationship to the ultimate truth of Christianity.

On its own, however, such internal consistency is only of limited

[1] This lack of ref. in Rahner to a Church without bishops is pointed out by Michael A. Fahey in Leo J. O'Donovan (ed.), 'A Changing Ecclesiology in a Changing Church: A Symposium on Development in the Ecclesiology of Karl Rahner', *TS* 38 (1977), 736–62.

[2] Gordon D. Kaufman, 'Is There Any Way from Athens to Jerusalem?', *Journal of Religion*, 59 (1979), 345.

value. Indeed, as was remarked in analysing Rahner's understanding of 'Church' in Chapter 1, he himself recognized the need for an external standard to overcome the danger of theologians' succumbing to their own idiosyncrasies. Since Rahner's proposals for change were designed to make the Church neither more cost-effective nor more popular but more unambiguously revelatory of the grace of God at work in history, the yardstick which is most suitable for evaluating his analysis is the notion of 'catholicity', a heading under which can be summarized the essential elements of the Church. The first step in applying this standard is to list what Avery Dulles considers to be constitutive of an authentic catholicity.

In Dulles's assessment, the most important aspect of catholicity is its multi-faceted nature. Since it designates the fullness of reality, catholicity actually demands differences in the Church, as only thus can the fullness of reality be represented. Such diversity, because it reflects divine life, is not, however, destructive:

> [catholicity] is a dynamic term. It designates a fullness of reality and life, especially divine life, actively communicating itself. This life, flowing outwards, pulsates through many subjects, draws them together, and brings them into union with their source and goal. By reason of its supreme realization, which is divine, catholicity assures the ultimate coherence of the whole ambit of creation and redemption.[3]

When applied to the Church, the dynamism implicit in catholicity suggests a Church which is not simply incomplete but which is questing after its own completeness. Thus, Dulles argues that catholicity is to be understood as both a gift and a task: a gift because of the presence in the Church—and in the variety of experiences of those who compose the Church—of the God who is the source of life, and a task because the Church is challenged to become more holy, more faithful, more united, and more deeply implanted in human culture if it is really to symbolize the presence of that God in history.[4] An authentic catholicity therefore requires what Dulles, following Tillich, refers to as a 'Catholic principle', which recognizes and protects the mediatory structures of the Church, and a 'Protestant principle', which stresses the immediacy of the Spirit animating all members of the Church and which prevents the structures themselves becoming a source of

[3] Avery Dulles, *The Catholicity of the Church* (Oxford, 1985), 167.
[4] Ibid. 170.

alienation.[5] Since the Spirit is active in all the various members of the Church, Dulles stresses that uniformity cannot be accepted as a genuine expression of the Church's catholicity. Consequently, the aim of a truly catholic Church is not uniformity, but a 'reconciled diversity'.[6]

A further element which Dulles identifies as essential to catholicity is the need for symbols—words and actions—which embody the Church's unity and which in so doing are accessible to all. Without such symbols, the unity of the Church would be based solely on the transcendental experience of grace. In the concrete, what is needed is a process of inculturation which both respects the Church's inherited symbols and is open to the experience and symbols of non-Europeans.[7] Furthermore, Dulles claims that catholicity not only requires an openness to the symbols of different cultures, but also involves respect for the 'partial catholicities' of philosophy, science, politics, economics, and the arts. The challenge for the Church is to see such disciplines as related to Christ, who is the centre and crown of all creation. Consequently, the Church can provide a 'vision of ultimacy' without needing to claim expertise in all the technical aspects of secular life.[8] For Dulles, it is the critical openness to all possible sources of knowledge which is characteristic of catholicity:

> The Catholic spirit, distinguished from docility, continues to cherish the wisdom handed down from earlier centuries. Self-effacing, it seeks to build on the work of others, and to contribute in modest but significant ways to a continuing tradition. This spirit, moreover, is prepared to learn from all parties, seeking out the truth in every opinion and the merit in every cause. . . . To be truly Catholic means to call into question the self-interest of any group, even that of the Church itself, and to maintain critical distance from every passing vogue. In short, catholicity seeks to foster a vision and concern as deep and comprehensive as God's redemptive plan, made known to us in his Son.[9]

When assessed in light of Dulles's analysis of catholicity, Rahner's project for change in the Church, as well as his general ecclesiology, emerges as unequivocally catholic. To support this conclusion, it is necessary to review the details of Rahner's approach: first, both

[5] Ibid. 169–70. For more on the 'Protestant principle', see Ronald Modras, *Paul Tillich's Theology of the Church* (Detroit, 1976), 106–28.

[6] Dulles, The Catholicity of the Church, 174.

[7] Ibid. 176–7.

[8] Ibid. 179.

[9] Ibid. 180.

his sacramental ecclesiology and his stress on the sinfulness of the Church meant that the dynamism of the Church, its growth in the direction of becoming the symbol of freedom in the world, was of prime importance to Rahner; secondly, his emphasis on freedom, the charismatic dimension, individual morality, and the local Church all contributed to promoting diversity in the Church; thirdly, while Rahner could thus be said to have reflected the 'Protestant principle', his commitment both to a structured Church and an authoritative teaching office showed he was no less aware of the 'Catholic principle'; fourthly, Rahner clearly recognized the virtues of the 'partial catholicities' of philosophy and the sciences. Hence, his work stressed not only the need for dialogue and for the Church to learn from the sciences, but also for the Church to acknowledge the authentic freedom of the sciences. At the same time, however, his recognition of the 'cosmic truth' to be found in Christ enabled him to challenge the limited vision of the sciences; fifthly, Rahner retained a breadth of vision which empowered him to criticize both the narrowness of office-holders who resisted all change, and the flawed liberalism of those prepared to separate the Church from its roots. In short, Rahner's commitment to the 'centre' embodied the comprehensiveness essential to catholicity.

If Rahner's theology can thus be commended as enhancing the Church's catholicity, that does not imply that he resolved every question that can be asked in regard to the Church's capacity to change. Indeed, one area where Rahner identified the issues without uttering the last word is the crucial one of the activity of the Spirit in the Church. As was pointed out on a number of occasions throughout this study, there was a tension in Rahner's work between his willingness to trust that the Spirit could preserve the Church from the evils of institutionalization and the abuse of power, and his desire that the rights of different groups in the Church, above all the laity, be given statutory protection. Consequently, his defence of the Pope's *Kompetenz der Kompetenz* was not always at ease with his desire to guard against excessive centralism.

In fact, the historical evidence of abuse of authority which Rahner himself adduced did little to support his conviction that the Spirit would protect the charisms of all groups in the Church. Accordingly, Rahner did not explain how the activity of the Spirit had been expressed either in the period when the feudal model of the episcopate made it impossible for the laity to take an active role in guiding the Church or during the Pian era, when the excessive centralization of authority and the poverty of theology meant that the Church had been in danger of extinguishing the Spirit. While Rahner's reliance on the Spirit to reconcile the potentially conflicting claims of the local and the universal Church, the laity and the hierarchy, and the episcopate and the primacy was an example of that hoping-against-hope which he regarded as characteristic of Christian faith, there nevertheless remains a need for a greater precision in the understanding of the relationship between human law and the activity of the Spirit in the Church.[10]

Another area where Rahner's work needs to be developed further concerns the mechanics of introducing alternative models of action in the Church. In suggesting changes in the Church's structures, Rahner often concentrated on a broad vision of what was possible, rather than on a detailed analysis of how such changes might move from idea to reality. Thus, for example, his proposals that bishops be elected and that the laity ought to have a greater say in the decisions reached by those in authority were slight on attention to practical application. That lack of detail was not necessarily a major deficiency, as it was consistent with Rahner's approach that such decisions be made on the local level rather than follow a model to be applied universally. Similarly, his ideas on experiment in the Church, which reflected his commitment to a Church where not everything flowed from the centre, did not exhaustively analyse the relationship between those experiments and preservation of the Church's unity and continuity. Since unity and continuity were regarded by Rahner as achievements of the Spirit alone, it would not have been possible for him to specify how they could be affected by experiments. Trust in the Spirit ought not exclude, however, ongoing reflection on such issues.

Even when Rahner was vague on the details of how changes might be applied, what nevertheless emerged with clarity from his work was his commitment to a Church where faithfulness to

[10] For criticism of Rahner's lack of a developed understanding of the work of the Spirit in the Church, see Felix Senn, *Orthopraktische Ekklesiologie? Karl Rahners Offenbarungsverständnis und seine ekklesiologischen Konsequenzen im Kontext der neueren katholischen Theologiegeschichte* (Fribourg, 1989), 580. See also the discussion of Senn's book in Karl-Heinz Neufeld's review in *ZKTh* 112 (1990), 355–7.

tradition did not imply the mere repetition of what had always been done. For Rahner, continuity and faithfulness to tradition were perfectly compatible with seeing the Church in a new way. By adopting such an approach, Rahner echoed Maurice Blondel, who advocated a dynamic understanding of tradition, an understanding which suggested that: 'Tradition's powers of conservation are equalled by its powers of conquest.'[11] While such a position might be unattractive both to those who oppose any change in the Church and to those who regard the past as simply a burden to be borne, Rahner's approach can nevertheless mediate between less extreme expressions of the conservative and progressive philosophies.

Once it is accepted that change, no less than preservation, is essential to tradition and the identity of the Church, then Vatican II can be regarded as developing, not abandoning, the Church's heritage. Similarly, Rahner's emphasis on the Spirit at work in history provides a rationale for a more positive asssessment of history than that adduced by those who equate secular society with decadence and godlessness. In encouraging the Church to learn from society, Rahner was not courting cheap popularity or naïvely ceding the Church's capacity to challenge destructive social trends. He was, however, making clear that the Spirit was not to be divorced from 'the times', even when those times were significantly different from the halcyon days of the Holy Roman Empire. In short, Rahner's approach allows substantial reform, including that of the magnitude introduced by Vatican II, to be interpreted positively rather than portrayed as an abject surrender to fashion. Consequently, we ought to expect the history of the Church to be a history of change.

His commitment to the new was, however, only one aspect of Rahner's approach. As has been continually emphasized in this work, Rahner also prized continuity in the Church as an expression of faithfulness to God's revelation in Christ. None the less, 'continuity' too was to be understood as a dynamic rather than a static term. As has been noted, Rahner believed that the unchangeable in the Church's life could only be discovered by an act of faith in the midst of change. Paradoxically, a commitment to

[11] Maurice Blondel, 'History and Dogma', in id., *The Letter on Apologetics and History and Dogma*, trans. A. Dru and I. Trethowan (London, 1964), 267. See also the discussion of Blondel in Dulles, *Reshaping of Catholicism*, 86–90.

continuity called for an openness to change. A balanced response demanded, therefore, that this openness be supplemented by a reverence for what had been inherited, a recognition that the Church was not to be constructed anew in every age. In other words, while Rahner was, in terms of the categories outlined in the Introduction, committed to reform in the Church, he was not proposing a reformation.

For those genuinely seeking a way between the extremes of a rigid conservatism and a facile progressivism, Rahner's approach has much to offer. He managed to combine an openness to change with a deep commitment to what forged the Church's identity. Moreover, he did so without abandoning the tools of theological analysis in favour of a purely sociological approach to the Church. Indeed, Rahner was able to unite both respect for the human sciences with a consistently Catholic theology. The result was an ecclesiology which both challenges and encourages, which esteems the past and offers hope for the future.

BIBLIOGRAPHY

The bibliography is divided into two sections: the first lists Rahner's own works; the second lists works by other authors.

In the Rahner section, the first part is devoted to articles in *Theological Investigations* and *Schriften zur Theologie*. The articles cited in each volume are those which have been referenced throughout this book. As will be evident, the content of the English and German volumes coincides only for the first six volumes; after that, the numbering of the English volumes differs substantially from the German originals. For each article, the date of the original publication is given in brackets after the English title.

For books written by Rahner, the details of the English version are followed by the title and details of the original German edition.

WORKS BY KARL RAHNER

Theological Investigations

VOLUME I: *God, Christ, Mary and Grace*, trans. Cornelius Ernst (New York, 1982). Originally published in 1961.

'The Development of Dogma', 39–77 (1954)

'Current Problems in Christology', 149–220 (1954)

'The Immaculate Conception', 201–13 (1954)

'Theological Reflections on Monogenism', 229–96 (1954)

'Concerning the Relationship between Nature and Grace', 297–317 (1950)

'Some Implications of the Scholastic Concept of Uncreated Grace', 319–46 (1939)

VOLUME II: *Man in the Church*, trans. Karl-Heinz Kruger (New York, 1975). Originally published in 1963.

'Membership of the Church According to the Teaching of Pius XII's Encyclical "Mystici Corporis Christi"', 1–88 (1947)

'Freedom in the Church', 89–107 (1953)

'Personal and Sacramental Piety', 109–33 (1952)

'Forgotten Truths Concerning the Sacrament of Penance', 153–74 (1953)

'On the Question of a Formal Existential Ethic', 217–34 (1955)

'The Dignity and Freedom of Man', 235–63 (1952)

'Peaceful Reflections on the Parochial Principle', 283–318 (1948)
'Notes on the Lay Apostolate', 319–52 (1955)

VOLUME III: *The Theology of the Spiritual Life*, trans. Karl-Heinz and Boniface Kruger (New York, 1982). Originally published in 1967.

'Thoughts on the Theology of Christmas', 24–34 (1955)
'The Eternal Significance of the Humanity of Jesus for our Relationship with God', 35–46 (1953)
'Reflections on the Experience of Grace', 86–90 (1954)
'The Church of the Saints', 91–104 (1955)
'The Dogma of the Immaculate Conception in our Spiritual Life', 129–40 (1954)
'The Meaning of Frequent Confessions of Devotion', 177–89 (1934)
'Priestly Existence', 239–62 (1942)
'The Consecration of the Layman to the Care of Souls', 263–76 (1936)
'The Christian among Unbelieving Relations', 355–72 (1954)
'On Conversions to the Church', 373–84 (1953)

VOLUME IV: *More Recent Writings*, trans. Kevin Smyth (New York, 1982). Originally published in 1966.

'Considerations on the Development of Dogma', 3–35 (1958)
'The Concept of Mystery in Catholic Theology', 36–73 (1959)
'On the Theology of the Incarnation', 105–20 (1958)
'Virginitas in Partu', 134–62 (1960)
'Nature and Grace', 165–88 (1959)
'The Theology of the Symbol', 221–52 (1959)

VOLUME V: *Later Writings*, trans. Karl-Heinz Kruger (New York, 1983). Originally published in 1966.

'Thoughts on the Possibility of Belief Today', 3–22 (1962)
'Theology in the New Testament', 23–41 (1962)
'What is a Dogmatic Statement?', 42–66 (1961)
'History of the World and Salvation History', 97–114 (1962)
'Christianity and the Non-Christian Religions', 115–34 (1961)
'Christianity and the "New Man"', 135–53 (1961)
'Christology within an Evolutionary View of the World', 157–92 (1962)
'Reflections on the Concept of "Ius Divinum" in Catholic Thought', 219–43 (1962)
'On the Theology of the Council', 244–67 (1962)
'The Theology of the Restoration of the Diaconate', 268–314 (1962)
'Some Remarks on the Question of Conversions', 314–35 (1962)
'Dogmatic Notes on "Ecclesiological Piety"', 336–65 (1961)
'Some Theses on Prayer "In the Name of the Church"', 419–38 (1961)
'What is Heresy?', 468–512 (1961)

VOLUME VI: *Concerning Vatican Council II*, trans. Karl-Heinz and Boniface Kruger (New York, 1982). Originally published in 1969.

'The Man of Today and Religion', 3–20 (1962)

'A Small Question Regarding the Contemporary Pluralism in the Intellectual Situation of Catholics and the Church', 21–30 (1965)

'Reflections on Dialogue within a Pluralist Society', 31–42 (1965)

'Ideology and Christianity', 43–58 (1965)

'Marxist Utopia and the Christian Future of Man', 59–68 (1965)

'Scripture and Tradition', 98–112 (1963)

'The Theology of Freedom', 178–96 (1964)

'Justified and Sinner at the Same Time', 218–30 (1963)

'The Church of Sinners', 253–69 (1947)

'The Sinful Church in the Decrees of Vatican II', 270–94 (1965)

'The Church and the Parousia of Christ', 295–312 (1963)

'The Episcopal Office', 313–60 (1963)

'Pastoral-Theological Observations on Episcopacy in the Teaching of Vatican II', 361–88 (1965)

'On Bishops' Conferences', 369–89 (1963)

VOLUME VII: *Further Theology of the Spiritual Life: 1*, trans. David Bourke (New York, 1977). Originally published in 1971.

'Christian Living Formerly and Today', 3–24 (1966)

'Do Not Stifle the Spirit!', 72–87 (1962)

'The Christian in his World', 88–99 (1965)

'"I Believe In The Church"', 100–18 (1954)

'The Church as the Subject of the Sending of the Spirit', 186–92 (1956)

VOLUME VIII: *Further Theology of the Spiritual Life: 2*, trans. David Bourke (New York, 1977). Originally published in 1971.

'The Sacramental Basis for the Role of the Layman in the Church', 51–74 (1960)

'The Position of Women in the New Situation in which the Church Finds Herself', 75–93 (1964)

'On The Situation of the Catholic Intellectual', 94–111 (1966)

VOLUME IX: *Writings of 1965–67: 1*, trans. Graham Harrison (New York, 1972). Originally published in 1972.

'The Second Vatican Council's Challenge to Theology', 3–27 (1966)

'Theology and Anthropology', 28–45 (1966)

'Philosophy and Philosophising in Theology', 46–83 (1968)

'The Historicity of Theology', 64–82 (1966)

'Theology and the Church's Teaching Authority after the Council', 83–100 (1966)

'Practical Theology within the Totality of Theological Disciplines', 101–13 (1967)

'The Need for a "Short Formula" of Christian Faith', 117–26 (1967)

VOLUME X: *Writings of 1965–67: 2*, trans. David Bourke (New York, 1977). Originally published in 1973.

'The New Image of the Church', 3–29 (1966)

'Church, Churches, and Religions', 30–49 (1966)
'On the Relationship between the Pope and the College of Bishops', 50–70 (1967)
'The Presence of the Lord in the Christian Community at Worship', 71–83 (1966)
'On the Presence of Christ in the Diaspora Community According to the Teaching of the Second Vatican Council', 84–102 (1966)
'Dialogue in the Church', 103–21 (1967)
'The Teaching of the Second Vatican Council on the Diaconate', 222–32 (1966)
'On the Theological Problems Entailed in a "Pastoral Constitution"', 293–317 (1967)
'Theological Reflections on the Problem of Secularisation', 318–48 (1967)
'Practical Theology and the Social Work of the Church', 349–70 (1967)
VOLUME XI: *Confrontations: 1*, trans. David Bourke (New York, 1982). Originally published in 1974.
'Pluralism in Theology and the Unity of the Creed in the Church', 3–23 (1969)
'On the Theology of the Ecumenical Discussion', 24–67 (1968)
'The New Claims which Pastoral Theology Makes upon Theology as a Whole', 115–36 (1969)
'Theological Considerations on Secularization and Atheism', 166–84 (1968)
'Reflections on the Problems Involved in Devising a Short Formula of the Faith', 230–44 (1970)
'On the Encyclical "Humanae Vitae"', 263–87 (1968)
VOLUME XII: *Confrontations: 2*, trans. David Bourke (New York, 1974). Originally published in 1974.
'The Teaching Office of the Church in the Present-Day Crisis of Authority', 3–30 (1970)
'The Point of Departure in Theology for Determining the Nature of the Priestly Office', 31–8 (1969)
'Theological Reflections on the Priestly Image of Today and Tomorrow', 39–60 (1969)
'On the Diaconate', 61–80 (1969)
'Observations on the Factor of the Charismatic in the Church', 81–97 (1969)
'Schism in the Catholic Church?', 98–116 (1969)
'Heresies in the Church Today?', 117–41 (1968)
'Concerning our Assent to the Church as She Exists in the Concrete', 142–60 (1969)
'Anonymous Christianity and the Missionary Task of the Church', 161–78 (1970)

'The Question of the Future', 181–201 (1969)
'Perspectives for the Future of the Church', 202–17 (1968)
'On the Structure of the People of the Church Today', 218–28 (1970)
'The Function of the Church as a Critic of Society', 229–49 (1969)

VOLUME XIII: *Theology, Anthropology, Christology*, trans. David Bourke (New York, 1983). Originally published in 1975.
'Institution and Freedom', 105–21 (1971)

VOLUME XIV: *Ecclesiology, Questions in the Church, the Church in the World*, trans. David Bourke (New York, 1976).
'Basic Observations on the Subject of the Changeable and Unchangeable in the Church', 3–23 (1972)
'The Faith of the Christian and the Doctrine of the Church', 24–46 (1972)
'Does the Church Offer any Ultimate Certainties?', 47–65 (1972)
'On the Concept of Infallibility in Catholic Theology', 66–84 (1970)
'The Dispute Concerning the Church's Teaching Office', 85–97 (1970)
'The Congregation of the Faith and the Commission of Theologians', 98–115 (1970)
'On the Theology of a "Pastoral Synod"', 116–31 (1970)
'Aspects of the Episcopal Office', 185–201 (1972)
'How the Priest Should View his Official Ministry', 202–19 (1972)
'Some Problems in Contemporary Ecumenism', 245–53 (1972)
'Ecumenical Theology in the Future', 254–69 (1972)
'The Unreadiness of the Church's Members to Accept Poverty', 270–9

VOLUME XVII: *Jesus, Man, and the Church*, trans. Margaret Kohl (New York, 1981).
'Opposition in the Church', 127–38 (1974)
'"Mysterium Ecclesiae"', 139–55 (1973)
'The Area Bishop: Some Theological Reflections', 151–66 (1973)
'Transformations in the Church and Secular Society', 167–80 (1975)
'The One Church and the Many Churches', 183–96 (1968)
'Is Church Unity Dogmatically Possible?', 197–214 (1973)
'Third Church?', 215–27 (1973)

VOLUME XVIII: *God and Revelation*, trans. Edward Quinn (New York, 1983).
'Magisterium and Theology', 54–73 (1978)

VOLUME XIX: *Faith and Ministry*, trans. Edward Quinn (New York, 1983).
'The Church's Redemptive Provenance from the Death and Resurrection of Jesus', 24–38 (1980)
'Consecration in the Life and Reflection of the Church', 57–62 (1980)
'Pastoral Ministries and Community Leadership', 73–86 (1977)
'Basic Communities', 159–65 (1980)

VOLUME XX: *Concern for the Church*, trans. Edward Quinn (New York, 1986). Originally published in 1981.

'Courage for an Ecclesial Christianity', 3–12 (1979)
'On the Situation of Faith', 13–32 (1980)
'Women and the Priesthood', 35–47 (1977)
'The Church's Responsibility for the Freedom of the Individual', 51–64 (1980)
'Basic Theological Interpretation of the Second Vatican Council', 77–89 (1979)
'The Abiding Significance of the Second Vatican Council', 90–102 (1979)
'The Future of the Church and the Church of the Future', 103–14 (1980)
'Structural Change in the Church of the Future', 115–32 (1977)
'Dream of the Church', 133–42 (1978)
'The Spirituality of the Church of the Future', 143–53 (1980)
'Unity of the Church—Unity of Mankind', 154–72 (1978)

VOLUME XXI: *Science and Theology*, trans. Hugh M. Riley (New York, 1988).

'Aspects of European Theology', 78–98 (1983)
'A Theology that We Can Live with', 99–112 (1983)
'The Act of Faith and the Content of Faith', 151–61 (1982)
'A Hierarchy of Truths', 162–7 (1982)

VOLUME XXII: *Humane Society and the Church of Tomorrow*, trans. Joseph Donceel (New York, 1991).

'Realistic Possibility of a Unification in Faith?', 67–79 (1983)
'Concrete Official Steps towards Unification', 80–3 (1982)
'Ecumenical Togetherness Today', 84–93 (1980)
'Perspectives for Pastoral Theology in the Future', 106–19 (1981)
'The Future of Christian Communities', 120–33 (1982)
'Rites Controversy: New Tasks for the Church', 134–9 (1983)
'South American Base Communities in a European Church', 148–54 (1981)
'Christian Pessimism', 155–62 (1984)
'What the Church Officially Teaches and What the People Actually Believe', 165–75 (1981)
'Theology and the Roman Magisterium', 176–90 (1980)
'The Perennial Actuality of the Papacy', 191–208 (1983)

Schriften zur Theologie

BAND I: *Gott, Christus, Maria, Gnade* (Cologne, 1967). Originally published in 1954.

'Zur Frage der Dogmenentwicklung', 49–90

'Probleme der Christologie von heute', 169–222
'Die unbefleckte Empfängnis', 223–37
'Theologisches zum Monogenismus', 253–322
'Über das Verhältnis von Natur und Gnade', 323–45
'Zur scholastischen Begrifflichkeit der ungeschaffenen Gnade', 347–75
BAND II: *Kirche und Mensch* (Cologne, 1968). Originally published in 1955.
'Die Gliedschaft in der Kirche nach der Lehre der Enzyklika Pius XII "Mystici Corporis Christi"', 7–94
'Die Freiheit in der Kirche', 95–114
'Personale und sakramentale Frömmigkeit', 115–41
'Vergessene Wahrheiten über das Bußsakrament', 143–83
'Über die Frage einer formalen Existentialkethik', 227–46
'Würde und Freiheit des Menschen', 247–77
'Friedliche Erwägungen über das Pfarrprinzip', 299–337
'Über das Laienapostolat', 339–73
BAND III: *Zur Theologie des Geistlichen Lebens* (Cologne, 1967). Originally published in 1956.
'Zur Theologie der Weihnachtsfeier', 35–46
'Die ewige Bedeutung der Menschheit Jesu für unser Gottesverhältnis', 47–60
'Über die Erfahrung der Gnade', 105–9
'Die Kirche der Heiligen', 111–26
'Das Dogma von der unbefleckten Empfängnis Mariens und unsere Frömmigkeit', 155–67
'Vom Sinn der häufigen Andachtsbeichte', 211–25
'Priesterliche Existenz', 285–312
'Weihe des Laien zur Seelsorge', 313–28
'Der Christ und seine ungläubigen Verwandten', 419–39
'Über Konversionen', 441–53
BAND IV: *Neuere Schriften* (Cologne, 1967). Originally published in 1960.
'Überlegungen zur Dogmenentwicklung', 11–50
'Zur Theologie der Menschwerdung', 137–55
'Virginitas in Partu', 173–205
'Natur und Gnade', 209–36
'Zur Theologie des Symbols', 275–311
BAND V: *Neuere Schriften* (Cologne, 1968). Originally published in 1962.
'Über die Möglichkeit des Glaubens heute', 11–32
'Theologie im Neuen Testament', 33–53
'Was ist eine dogmatische Aussage?', 54–81
'Weltgeschichte und Heilsgeschichte', 115–35
'Das Christentum und die nichtchristlichen Religionen', 136–58
'Das Christentum und der "Neue Mann"', 159–79
'Die Christologie innerhalb einer evolutiven Weltanschauung', 183–221

'Über den Begriff "Ius Divinum" im katholischen Verständnis', 249–77
'Zur Theologie des Konzils', 278–302
'Die Theologie der Erneuerung des Diakonates', 303–55
'Einige Bemerkungen über die Frage der Konversionen', 356–78
'Dogmatische Randbemerkungen zur "Kirchenfrömmigkeit"', 379–410
'Thesen über das Gebet "im Namen der Kirche"', 471–93
'Was ist Häresie?', 527–76

BAND VI: *Neuere Schriften* (Cologne, 1968). Originally published in 1965.
'Der Mensch von heute und die Religion', 13–33
'Kleine Frage zum heutigen Pluralismus in der geistigen Situation der Katholiken und der Kirche', 34–95
'Über den Dialog in der pluralistischen Gesellschaft', 46–58
'Ideologie und Christentum', 59–76
'Marxistische Utopie und christliche Zukunft des Menschen', 77–88
'Heilige Schrift und Tradition', 121–38
'Theologie der Freiheit', 215–37
'Gerecht und Sünder zugleich', 262–76
'Kirche der Sünder', 301–20
'Sündige Kirche nach den Dekreten des zweiten Vatikanischen Konzils', 321–47
'Kirche und Parousie Christi', 348–67
'Über den Episkopat', 369–422
'Pastoraltheologische Bemerkungen über den Episkopat in der Lehre des II Vatikanum', 423–31
'Über Bischofkonferenzen', 432–54
'Kirche im Wandel', 455–78
'Konziliare Lehre der Kirche und künftige Wirklichkeit christlichen Lebens', 479–98
'Grenzen der Amtskirche', 499–520
'Der Anspruch Gottes und der Einzelne', 521–36
'Zur "Situationsethik" aus ökumenischer Sicht', 537–44

BAND VII: *Zur Theologie des Geistlichen Lebens* (Cologne, 1971). Originally published in 1966.
'Frömmigkeit früher und heute', 11–31
'Löscht den Geist nicht aus', 77–90
'Der Christ in seiner Umwelt', 91–102
'Ich glaube die Kirche', 103–20
'Die Kirche als Ort der Geistsendung', 183–8
'Sakramentale Grundlegung des Laienstandes in der Kirche', 330–50
'Die Frau in der neuen Situation der Kirche', 351–67
'Zur Situation des katholischen Intellektuellen', 368–85

BAND VIII: *Theologische Vorträge und Abhandlungen* (Cologne, 1967).
'Die Herausforderung der Theologie durch das Zweite Vatikanische Konzil', 13–42

'Theologie und Anthropologie', 43–65
'Philosophie und Philosophieren in der Theologie', 66–87
'Zur Geschlichkeit der Theologie', 88–110
'Kirchliches Lehramt und Theologie nach dem Konzil', 111–32
'Die praktische Theologie im Ganzen der theologischen Disziplinen', 133–49
'Die Forderung nach einer "Kurzformel" des christlichen Glaubens', 153–64
'Das neue Bild der Kirche', 329–54
'Kirche, Kirchen und Religionen', 355–73
'Zum Verhältnis zwischen Papst und Bischofskollegium', 374–94
'Die Gegenwart des Herrn in der christlichen Kultgemeinde', 395–408
'Über die Gegenwart Christi in der Diasporagemeinde nach der Lehre des Zweiten Vatikanischen Konzils', 409–25
'Vom Dialog in der Kirche', 426–44
'Die Lehre des Zweiten Vatikanischen Konzils über den Diakonat', 541–52
'Zur theologischen Problematik einer "Pastoralkonstitution"', 616–36
'Theologische Reflexionen zur Säkularisation', 637–66
'Praktische Theologie und kirchliche Sozialarbeit', 667–88

BAND IX: *Konfrontationen* (Cologne, 1970).
'Der Pluralismus in der Theologie und die Einheit des Bekenntnisses in der Kirche', 11–33
'Zur Theologie des ökumenischen Gesprächs', 34–78
'Neue Ansprüche der Pastoraltheologie an die Theologie als ganze', 127–47
'Theologische Überlegungen zu Säkularisation und Atheismus', 177–96
'Reflexionen zur Problematik einer Kurzformel des Glaubens', 242–56
'Zur Enzyklika "Humanae Vitae"', 276–301
'Das kirchliche Lehramt in der heutigen Autoritätskrise', 339–65
'Der theologische Ansatzpunkt für die Bestimmung des Wesens des Amtspriestertums', 366–72
'Theologische Reflexionen zum Priesterbild von heute und morgen', 373–94
'Über den Diakonat', 395–414
'Bemerkungen über das Charismatische in der Kirche', 415–31
'Schisma in der katholischen Kirche?', 432–52
'Häresien in der Kirche heute?', 453–78
'Über das Ja zur konkreten Kirche', 479–97
'Anonymes Christentum und Missionsauftrag der Kirche', 498–515
'Die Frage nach der Zukunft', 519–40
'Perspektiven für die Zukunft der Kirche', 541–57
'Zur Struktur des Kirchenvolkes heute', 558–68
'Die gesellschaftkritische Funktion der Kirche', 569–90

BAND X: *Im Gespräch mit der Zukunft* (Cologne, 1972).
'Institution und Freiheit', 115–32
'Grundsätzliche Bemerkungen zum Thema: Wandelbares und Unwandelbares in der Kirche', 241–61
'Der Glaube des Christens und die Lehre der Kirche', 262–85
'Bietet die kirche letzte Gewiβheiten?', 286–304
'Zum Begriff der Unfehlbarkeit in der katholischen Theologie', 305–23
'Disput um das kirchliche Lehramt', 324–37
'Glaubenskongregation und Theologenkommission', 338–57
'Zur Theologie einer 'Pastoralsynode"', 358–73
'Aspekte des Bischofamtes', 430–47
'Zum Selbstverständnis des Amtspriesters', 448–66
'Einige Probleme des Ökumenismus heute', 493–502
Ökumenische Theologie in der Zukunft', 503–19
'Die Unfähigkeit zur Armut in der Kirche', 520–30
BAND XII: *Theologie aus Erfahrung des Geistes* (Cologne, 1975).
'Opposition in der Kirche', 469–81
'Mysterium Ecclesiae', 482–500
'Theologisches zur Aufgabe des Regionalbischofs', 501–12
'Kirchliche Wandlungen und Profangesellschaft', 513–28
'Die eine Kirche und die vielen Kirchen', 531–46
'Ist Kircheneinigung dogmatisch möglich?', 547–67
'Dritte Konfession?', 568–81
BAND XIII: *Gott und Offenbarung* (Cologne, 1978).
'Lehramt und Theologie', 69–92
BAND XIV: *In Sorge um die Kirche* (Cologne, 1980).
'Vom Mut zum kirchlichen Glauben', 11–22
'Zur Situation des Glaubens', 23–47
'Heilsgeschtliche Herkunft der Kirche von Tod und Auferstehung Jesu', 73–90
'Weihe im Leben und in der Reflexion der Kirche', 113–31
'Pastorale Dienste und Gemeindeleitung', 132–47
'Basisgemeinden', 265–72
'Priestertum der Frau?', 208–23
'Die Verantwortung der Kirche für die Frieheit des einzelnen', 248–64
'Theologische Grundinterpretation des II Vatikanischen Konzils', 287–302
'Die bleibende Bedeutung des II Vatikanischen Konzils', 303–18
'Die Zukunft der Kirche und die Kirche der Zukunft', 319–32
'Strukturwandel der Kirche in der künftigen Gesellschaft', 333–54
'Der Traum von der Kirche', 355–67

'Elemente der Spiritualität in der Kirche der Zukunft', 368–81
'Einheit der Kirche—Einheit der Menschheit', 382–404
BAND XV: *Wissenschaft und christlicher Glaube* (Cologne, 1983).
'Aspekte europäischer Theologie', 84–103
'Eine Theologie, mit der wir leben können', 104–16
'Glaubensakt und Glaubensinhalt', 152–62
'Hierarchie der Wahrheiten', 163–8
BAND XVI: *Humane Gesellschaft und Kirche von morgen* (Cologne, 1984).
'Realistische Möglichkeit der Glaubenseinigung?', 93–109 (1983)
'Konkrete offizielle Schritte auf eine Einigung hin?', 110–14 (1982)
'Ökumenisches Miteinander heute', 115–27 (1980)
'Perspektiven der Pastoral in der Zukunft', 143–59 (1981)
'Über die Zukunft der Gemeinden', 160–77 (1982)
'Ritenstreit: Neue Aufgaben für die Kirche', 178–84 (1983)
'Südamerikanische Basisgemeinden in einer europäischen Kirche', 196–205 (1981)
'Christlicher Pessimismus', 206–14 (1984)
'Offizielle Glaubenslehre der Kirche und faktische Gläubigkeit des Volkes', 217–30 (1981)
'Die Theologie und das römische Lehramt', 231–48 (1980)
'Die unvergängliche Akualität des Papsttums', 249–70

Other Works by Rahner

Books

'Blick in der Zukunft', in N. Greinacher and H. Risse (eds.), *Bilanz des deutschen Katholizismus* (Mainz, 1966).

The Christian of the Future, trans. W. J. O'Hara (London, 1967). These are articles from *ST* vi which were never included in *TI*.

The Church after the Council, trans. D. C. Herron and R. Albrecht (New York, 1966) (*Das Konzil: Ein neuer Beginn* (Freiburg i.B., 1966)).

The Church and the Sacraments, trans. W. J. O'Hara (London, 1963) (*Kirche und Sakramente* (Freiburg i.B., 1960)).

The Dynamic Element in the Church, trans. W. J. O'Hara (London, 1964) (*Das Dynamische in der Kirche* (Freiburg i.B., 1958)).

Faith In a Wintry Season: Conversations and Interviews with Karl Rahner in the Last Years of his Life, ed. H. Egan (New York, 1990).

(*Glaube in winterlicher Zeit: Gespräche mit Karl Rahner aus den letzten Lebensjahren*, ed. P. Imhof and H. Biallowons (Düsseldorf, 1986)).

Foundations of Christian Faith, trans. W. Dych (New York, 1978) (*Grundkurs des Glaubens: Einführung in den Begriff des Christentums* (Freiburg i.B., 1976)).

Free Speech in the Church, trans. G. R. Lamb (New York, 1959) (*Das freie Wort in der Kirche* (Einsiedeln, 1953)).

Freiheit und Manipulation in Gesellschaft und Kirche (Munich, 1970).

Grace In Freedom, trans. H. Graef (New York, 1969) (*Gnade als Freiheit* (Freiburg i.B., 1968)).

Hearers of The Word, rev. edn. Johann B. Metz, trans. M. Richards (Montreal, 1969) (*Hörer des Wortes: Zur Grundlegung einer Religionsphilosophie* 2nd. edn. (Freiburg i.B., 1971)).

I Remember, trans. H. Egan (London, 1985) (*Erinnerungen* (Freiburg i.B., 1984)).

Inspiration in the Bible, trans. H. Henkey (rev. trans. M. Palmer) (New York, 1964) (*Über die Schriftinspiration* (Freiburg i.B., 1958)).

Karl Rahner in Dialogue, trans. and ed. H. Egan (New York, 1986) (*Karl Rahner im Gespräch* ed. P. Imhof and H. Biallowons (2 vols.; Munich, 1983)).

Kritisches Wort (Freiburg i.B., 1970).

Meditations on Freedom and the Spirit, trans. R. Ockenden, D. Smith, and C. Bennett (London, 1977) (a compendium of *Freiheit und Manipulation in Gesellschaft und Kirche* (Munich, 1970) and *Toleranz in der Kirche* (Freiburg i.B., 1977)).

Mission and Grace, i, trans. C. Hastings (London, 1963) (*Sendung und Gnade* (Innsbruck, 1959)).

Nature and Grace, trans. D. Wharton (London, 1963) (*Gefahren im heutigen Katholizismus* (Einsiedeln, 1955)).

On Heresy, trans. W. J. O'Hara (London, 1964) ('Was ist Häresie?', in A. Böhm (ed.), *Häresien der Zeit* (Freiburg i.B., 1961)).

Opportunities for Faith, trans. E. Quinn (London, 1974) (*Chancen des Glaubens* (Freiburg i.B., 1971)).

'Selbstbesinnunung der Kirche', in J. C. Hampe (ed.), *Ende der Gegenreformation?* (Mainz, 1964).

The Shape of the Church to Come, trans. E. Quinn (London, 1974) (*Strukturwandel der Kirche als Chance und Aufgabe* (Freiburg i.B., 1989)).

Spirit in The World, trans. W. Dych (New York, 1968) (*Geist in Welt*, 2nd edn. (Munich, 1957)).

Theology of Pastoral Action, trans. W. J. O'Hara (Freiburg i.B., 1968). A translation of *Handbuch der Pastoraltheologie*, i.

'Theology of the Parish', in H. Rahner (ed.), *The Parish, From Theology to Practice* trans. R. Kress (Westminster, Md. 1958) (id. (ed.), *Die Pfarre: Von der Theologie zur Praxis* (Freiburg i.B., 1956)).

Toleranz in der Kirche (Freiburg i.B., 1977).

Vorfragen zu einem ökumenischen Amtsverständnis (Freiburg i.B., 1974).

(ed.), *Zum Problem Unfehlbarkeit: Antworten auf die Anfrage von Hans Küng* (Freiburg i.B., 1971).

Zur Reform des Theologiestudiums (Freiburg i.B., 1969).

Articles

'A Critique of Hans Küng: Concerning the Infallibility of Theological Propositions', *Homiletic and Pastoral Review*, 71 (May 1971), 10–26.

(with Hans Küng) 'A "Working Agreement" to Disagree', *America*, 129 (7 July, 1973), 11–12.

'Die zweite Konzilsperiode', *Oberrheinisches Pastoralblatt*, 65 (1964), 68–82.

'Eine neue Epoche in der Kirche eingeleitet', *Der Volksbote* (8 June, 1963), 2.

'Ich protestiere', *Publik-Forum*, 23 (1979), 15–19.

'Reply to Hans Küng: In the Form of an Apologia Pro Theologia Sua', *Homiletic and Pastoral Review*, 71 (Aug./Sept. 1971), 11–27.

'Wagnis oder Trägheit? Die Kirche und die geistige Situation der Gegenwart', *Universitas*, 18 (1963), 1209–15.

'Zu: "Warum schweigt Rahner?"', *Publik-Forum*, 25 Jan. (1980), 32.

Unpublished Manuscript

'Vienna Memorandum' (1943). Located in the Karl-Rahner-Archiv, Innsbruck.

WORKS CO-AUTHORED BY RAHNER

Die Antwort der Theologen (with Johannes B. Metz; Düsseldorf, 1968).

Diakonia in Christo (with Herbert Vorgrimler; Freiburg i.B., 1962).

Dictionary of Theology (with Herbert Vorgrimler; trans. R. Strachan, D. Smith, R. Nowell, and S. O'Brien Twohig (Freiburg i.B., 1976)).

The Episcopate and the Primacy (with Joseph Ratzinger; trans. K. Barker, P. Kerans, R. Ochs, and R. Strachan; New York, 1962) (*Episkopat und Primat* (Freiburg i.B., 1961)).

Handbuch der Pastoraltheologie (with Franz X. Arnold, V. Schurr, and L. Weber; 5 vols.; Freiburg i.B., 1964–9).

Kleines Konzilkompendium (with Herbert Vorgrimler; Freiburg i.B., 1966).

Revelation and Tradition (with Joseph Ratzinger; trans. W. J. O'Hara; Freiburg i.B., 1966) (*Offenbarung and Überlieferung* (Freiburg i.B., 1965)).

Sind die Erwartungen erfüllt? (with Oscar Cullman and Heinrich Fries (eds.); Munich, 1966).

Unity of the Churches: An Actual Possibility (with Heinrich Fries; trans. R. C. L. Gritsch and E. W. Gritsch (New York, 1985) (*Einigung der Kirchen: Reale Möglichkeit* (Freiburg i.B., 1987)).

WORKS BY OTHER AUTHORS

Books

ALTMANN, WALTHER, *Der Begriff der Tradition bei Karl Rahner* (Berne, 1974).

BACIK, JAMES J., *Apologetics and the Eclipse of Mystery: Mystagogy According to Karl Rahner* (Notre Dame, Ind., 1980).

BLONDEL, MAURICE, *Letter on Apologetics and History and Dogma*, trans. A. Dru and I. Trethowan (London, 1964).

BOFF, LEONARDO, *Die Kirche als Sakrament* (Paderborn, 1972).

BONSOR, JACK ARTHUR, *Rahner, Heidegger, and Truth: Karl Rahner's Notion of Christian Truth—The Influence of Martin Heidegger* (Lanham, 1987).

CALLAHAN, ANNICE, *Karl Rahner's Spirituality of the Pierced Heart: A Reinterpretation of Devotion to the Sacred Heart* (Lanham, 1985).

DENZIGER, HEINRICH, and SCHÖNMETZER, ADOLF (eds.), *Enchiridion Symbolorum* (36th edn.; Freiburg i.B., 1976).

DUFFY, STEPHEN J., *The Graced Horizon: Nature and Grace in Modern Catholic Thought* (Collegeville, 1992).

DULLAART, LEO, *Kirche und Ekklesiologie* (Munich, 1975).

DULLES, AVERY, *Models of the Church* (New York, 1978).

—— *The Resilient Church: The Necessity and Limits of Adaption* (Dublin, 1978).

—— *A Church To Believe in: Discipleship and the Dynamics of Freedom* (New York, 1982).

—— *Models of Revelation* (New York, 1983).

—— *The Catholicity of the Church* (Oxford, 1985).

—— *The Reshaping of Catholicism: Current Challenges in the Theology of Church* (San Francisco, 1988).

DYCH, WILLIAM V., *Karl Rahner* (London, 1992).

FLANNERY, AUSTIN (ed.), *The Documents of Vatican II* (New York, 1975).

—— *Vatican Council II*, ii (New York, 1982).

GREINACHER, NORBERT, and HAAG, HERBERT (eds.), *Der Fall Küng* (Munich, 1980).

HAIGHT, ROGER, *The Experience and Language of Grace* (Dublin, 1979).

HEBBLETHWAITE, PETER, *The Runaway Church* (New York, 1975).

HINES, MARY E., *The Transformation of Dogma: An Introduction to Karl Rahner on Doctrine* (New York, 1989).

KEHL, MEDARD, *Kirche als Institution: Zur theologischen Begründung des institutionellen Charakters der Kirche in der neuen deutschprachigen katholischen Ekklesiologie* (Frankfurt am Main, 1976).

KELLY, GEFFREY B. (ed.), *Karl Rahner: Theologian of the Graced Search for Meaning* (Minneapolis, 1992).

KERN, WALTER, *Außerhalb der Kirche kein Heil?* (Freiburg i.B., 1979).

KLINGER, ELMAR, and WITTSTADT, KLAUS (eds.), *Glaube im Prozess* (Freiburg i.B., 1984).

KRESS, ROBERT, *A Rahner Handbook* (Atlanta, 1982).

—— *The Church: Communion, Sacrament, Communication* (New York, 1985).

KÜNG, HANS, *The Church*, trans. R. and R. Ockenden (London, 1967).

—— *Truthfulness: The Future of The Church* (London, 1968).
—— *Infallible? An Enquiry*, trans. E. Mosbacher (London, 1971).
—— *Theology for the Third Millennium*, trans. P. Heinegg (New York, 1988).
—— *Reforming the Church Today*, trans. P. Heinegg, F. McDonagh, J. Maxwell, E. Quinn, A. Snidler (New York, 1990).
—— and TRACY, DAVID (eds.), *Paradigm Change in Theology*, trans. M. Kohl (Edinburgh, 1989).
MACAULAY, THOMAS B., *Critical and Historical Essays Contributed to the Edinburgh Review* (London, 1870).
MCCOOL, GERALD A., *Catholic Theology in the Nineteenth Century: The Quest for a Unitary Method* (New York, 1977).
—— (ed.), *A Rahner Reader* (New York, 1981).
MARSHALL, BRUCE, *Christology in Conflict: The Identity of a Saviour in Rahner and Barth* (Oxford, 1987).
METZ, JOHANN BAPTIST (ed.), *Gott in Welt*, ii (Freiburg i.B., 1964).
MODRAS, RONALD, *Paul Tillich's Theology of the Church* (Detroit, 1976).
MUCK, OTTO, *The Transcendental Method*, trans. W. D. Seidensticker (New York, 1968).
MUGGERIDGE, ANNE ROCHE, *The Desolate City: Revolution in the Catholic Church*, rev. edn. (San Francisco, 1990).
NEUMANN, KARL, *Der Praxisbezug der Theologie bei Karl Rahner* (Freiburg i.B., 1980).
NEUNER, JOSEF and ROOS, HEINRICH (eds.), *Der Glaube der Kirche in den Urkunden der Lehrverkündigung*, 12th edn. (Regensburg, 1971).
NEWMAN, JOHN H., *An Essay on the Development of Christian Doctrine* (Westminster, 1968).
—— *Apologia Pro Vita Sua* (London, 1976).
—— *An Essay in Aid of a Grammar of Assent* (Notre Dame, Ind., 1979).
NICHOLS, AIDAN, *From Newman To Congar: The Idea of Doctrinal Development from the Victorians to the Second Vatican Council* (Edinburgh, 1990).
NOWELL, ROBERT, *A Passion for Truth: Hans Küng—A Biography* (London, 1981).
O'DONOVAN, LEO J., *A World of Grace: An Introduction to the Themes and Foundations of Karl Rahner's Theology* (New York, 1981).
O'LEARY, JOSEPH, *Questioning Back: The Overcoming of Metaphysics in Christian Tradition* (Minneapolis, 1985).
O'MALLEY, JOHN W., *Tradition and Transitions: Historical Perspectives on Vatican II* (Wilmington, 1989).
PRZEWOZNY, BERNARD, *The Church as the Sacrament of the Unity of All Mankind* (Rome, 1979).
RICHARD, ROBERT L., *Secularization Theology* (New York, 1967).
ROBINSON, JOHN A. T., *Honest to God* (London, 1963).

Rynne, Xavier, *Vatican Council II* (New York, 1968).

Schnell, Ursula, *Das Verhältnis von Amt und Gemeinde im neueren Katholizismus* (Berlin, 1977).

Schwerdtfeger, Nikolaus, *Gnade und Welt: Zum Grundgefüge von Karl Rahners Theorie der 'anonymen Christen'* (Freiburg i.B., 1982).

Senn, Felix, *Orthopraktische Ekklesiologie? Karl Rahners Offenbarungsverständnis und seine ekklesiologischen Konsequenzen im Kontext der neueren katholischen Theologiegeschichte* (Fribourg, 1989).

Stacpoole, Alberic (ed.), *Vatican II Revisited by Those Who Were There* (Minneapolis, 1986).

Sullivan, Francis A., *Salvation outside the Church? Tracing the History of the Catholic Response* (New York, 1992).

Vass, George, *Understanding Karl Rahner*, i: *A Theologian in Search of a Philosophy* (London, 1985); ii: *The Mystery of Man and the Foundations of a Theological System* (London, 1985).

Vorgrimler, Herbert, *Understanding Karl Rahner: An Introduction to his Life and Thought,* trans. J. Bowden (New York, 1986).

—— (ed.), *Commentary on the Documents of Vatican II*, i, trans. K. Smyth (London, 1967).

—— (ed.), *Wagnis Theologie: Erfahrungen mit der Theologie Karl Rahners* (Freiburg i.B., 1979).

—— (ed.), *Sehnsucht nach dem geheimnisvollen Gott* (Freiburg i.B., 1990).

Walsh, Michael J., *The Heart of Christ in the Writings of Karl Rahner: An Investigation of its Christological Foundation as an Example of the Relationship between Theology and Spirituality* (Rome, 1977).

Wong, Joseph H. P., *Logos-Symbol in the Christology of Karl Rahner* (Rome, 1984).

Articles

Bresnahan, James F., 'Rahner's Ethics: Critical Natural Law in Relation to Contemporary Ethical Methodology', *Journal of Religion*, 56 (1976), 36–60.

Buckley, James, 'On Being A Symbol: An Appraisal of Karl Rahner', *TS* 40 (1979), 453–73.

Callahan, C. Annice, 'Karl Rahner's Theology of the Symbol: Basis for his Theology of the Church and the Sacraments', *Irish Theological Quarterly*, 49 (1982), 195–205.

Clasby, Nancy, 'Dancing Sophia: Rahner's Theology of Symbols', *Religion and Literature*, 25 (1993), 51–65.

Dulles, Avery, 'Ecclesial Futurology: Moving Towards the 1990s', *Proceedings of the Canon Law Society of America*, 47 (1986), 1–15.

—— 'A Half Century of Ecclesiology', *TS* 50 (1989), 419–42.

FAHEY, MICHAEL A., 'Continuity in the Church amid Structural Change', *TS* 35 (1974), 415–40.

FICHTER, JOSEPH H., 'The Church: Looking to the Future', *America*, 160 (1989), 189–92.

FRIES, HEINRICH, 'Theologische Methode bei John Henry Newman und Karl Rahner', *Catholica*, 33 (1979), 109–33.

HUGHES, JOHN J., 'Infallible? An Inquiry Considered, *TS* 32 (1971), 183–207.

HURD, ROBERT L., 'The Concept of Freedom in Rahner', *Listening*, 17 (1982), 138–52.

KAUFMAN, GORDON D., 'Is There Any Way from Athens to Jerusalem?', *Journal of Religion*, 59 (1979), 340–57.

KÜNG, HANS 'Im Interesse der Sache: Antwort an Karl Rahner', *Stimmen der Zeit*, 187 (1971), 43–64; 105–22.

LYNCH, PATRICK J., 'Secularization Affirms the Sacred: Karl Rahner', *Thought*, 61 (1986), 381–96.

MAAS-EWARD, THEODOR, 'Odo Casel OSB und Karl Rahner SJ: Disput über das Wiener Memorandum "Theologische und philosophische Zeitfragen im katholishcen deutschen Raum"', *Archiv für Liturgiewissenschaft*, 28 (1986), 193–234.

MCDERMOTT, JOHN M., 'The Christologies of Karl Rahner', *Gregorianum*, 67 (1986), 87–123; 297–327.

MODRAS, RONALD, 'Implications of Rahner's Anthropology for Fundamental Moral Theology', *Horizons*, 12 (1985), 70–90.

MOLNAR, PAUL D., 'Can We Know God Directly? Rahner's Solution from Experience', *TS* 46 (1985), 228–61.

NEUFELD, KARL-HEINZ, 'Theologen und Konzil: Karl Rahners Beitrag zum Zweiten Vatikanischen Konzil', *Stimmen der Zeit*, 202 (1984), 156–66.

—— 'Lehramtliche Mißvertändnisse: Zu Schwierigkeiten Karl Rahners in Rom', *ZKTh* 111 (1989), 420–30.

NEUMANN, KARL, 'Diaspora Kirche als *sacramentum mundi*: Karl Rahner und die Diskussion um Volkskirche-Gemeindekirche', *Trierer theologische Zeitschrift*, 97 (1982), 52–71.

NICHOLS, AIDAN, '"Einigung der Kirchen": An Ecumenical Controversy', *One in Christ*, 21 (1985), 139–66.

O'DONOVAN, LEO J. (ed.), 'A Changing Ecclesiology in a Changing Church: A Symposium on Development in the Ecclesiology of Karl Rahner', *TS* 38 (1977), 736–62.

—— 'Orthopraxis and Theological Method in Karl Rahner', *Catholic Theological Society of America: Proceedings*, 35 (1980), 47–65.

O'MALLEY, JOHN W., 'Historical Thought and the Reform Crisis of the Early Sixteenth Century', *TS* 28 (1967), 531–48.

—— 'Reform, Historical Consciousness and Vatican II's Aggiornamento', *TS* 32 (1971), 573–601.

O'MALLEY, JOHN W., 'Developments, Reforms, and Two Great Reformations', *TS* 44 (1983), 373–406.
O'MEARA, THOMAS F., 'Karl Rahner: Some Audiences and Sources for his Theology', *Communio*, 18 (1991), 237–58.
RATZINGER, JOSEPH, 'Luther und die Einheit der Kirchen', *Internationale katholische Zeitschrift*, 12 (1983), 568–82.
SOBRINO, JON, 'Karl Rahner and Liberation Theology', *Theology Digest*, 32 (1985), 257–60.
VANDERVELDE, GEORGE, 'The Grammar of Grace: Karl Rahner as a Watershed in Contemporary Theology', *TS* 49 (1988), 445–59.
VOGEL, ARTHUR A., 'H. Fries and K. Rahner's "The Unity of the Churches": Three Responses', *Ecumenical Trends*, 14 (1985), 97–102.

Unpublished Manuscript

ROSA, PAUL A. de, 'Karl Rahner's Concept of "Vorgriff": An Examination of its Philosophical Background' (Univ. of Oxford Ph.D. thesis, 1988).

INDEX